The Story of Fairbank Oil:
The Four Generations of the Family Producing Oil Longer than Anyone in the World.National Library of Canada Cataloguing in Publication
Includes bibliographic references and index.
ISBN 0-9733581-0-6
1. Fairbank Oil Properties Ltd.--History. 2. Fairbank family.I.Title.
HD9574.C24F33 2003 338.7'6223382'0971327 C2003-905982-0
With gratitude we acknowledge assistance in printing from The Imperial Oil Foundation and The Lambton County Historical Association. Connie Bell, manager of The Oil Museum of Canada in Oil Springs, is thanked for her help in smoothing the way for this book to be printed.
Cover photograph by Phil Hein
Design Consultants: Scott and Sally Martyn

Printed and bound in Canada by
Haines Frontier Printing Limited
Sarnia, Ontario
Second printing 2013

The Story of Fairbank Oil

Four Generations of the Family Producing Oil Longer Than Anyone in the World

By Patricia McGee

WORDS UNLIMITED INK.

Haines Frontier Printing Limited
SARNIA, ONTARIO

To Emory

for his illuminating energy

Contents

Introduction to the Four Generations of Fairbanks

First Generation
John Henry Fairbank, 1831–1914
Edna Crysler 1828–1894

Second Generation
Dr. Charles Oliver Fairbank 1858–1925
Clara Sussex 1877–1956

Third Generation
Charles Oliver Fairbank II (Charles Sr.) 1904–1982
Jean Harwood 1914–1998

Fourth Generation
Charles Oliver Fairbank III (Charlie) 1941–
Patricia McGee 1955–

Acknowledgements

There was a treasure trove of information in those truckloads of Fairbank papers.

The meticulous historical research by Edward (Ted) Phelps has largely gone unsung. Early in the 1960s, he was approached to tackle the enormous task of organizing the mountain of Fairbank papers stored in the turret of the family's Petrolia mansion and elsewhere. The house was being sold and the problem had to be addressed. Though there was a treasure trove of information in those truckloads of paper, he was the one who sorted, filed, labelled and boxed it so that it was finally made accessible.

Ted was given access to letters, records and diaries dating back to the 1850s to write his 1965 master thesis *John Henry Fairbank of Petrolia* (1831-1914) *A Canadian Entrepreneur.* His curiosity led him to read extensively all manner of texts, reports, newspapers, surveys, diaries and maps. He also fleshed out stories by interviewing a wide array of people. While writing his thesis, he was an assistant at the Lawson Memorial Library at the University of Western Ontario in London. Earlier, he earned a library degree at McGill University and had an honours Bachelor of Arts degree in history and English. He went on to become the head of the regional history library at the University of Western Ontario.

A wealth of knowledge has also come to light thanks to Col. Bruce Harkness who spent three decades as a gas commissioner for the Ontario Department of Mines. He compiled all his findings in a manuscript that was never published and it was Ted who drew attention to it. Our understanding of early oil development has been vastly enriched by his research and writing, yet Harkness too has gone largely unrecognized. In an effort to correct this oversight, the Ontario Petroleum Institute posthumously awarded him the Award of Merit in 1991.

Robert Cochrane, a petroleum geologist with Cairnlins Resources Limited, offered his expertise in all technical matters and has great historical knowledge as well. For more than 20 years, he had dedicated himself in a myriad of ways to The Petrolia Discovery where he served as chairman for many years. Claudia Cochrane, also of Cairnlins Resources Limited, shared her expertise as a geologist. When asked why there was oil in Oil Springs, she wrote a succinct summary that compressed 375 million years of geology into a small, tidy package.

Dr. Emory Kemp, of the West Virginia University, deserves special thanks. In bringing his team of industrial archaeologists to the site, he brought fresh enthusiasm and is laying the foundation for future historical designations. This book began because Dr. Kemp said he'd like "a little background information" for the industrial archaeologists.

Charlie Fairbank has made the recognition of the early oil industry his true life's work. It was a proud moment in 1997 when he accepted the award to J.H. Fairbank in the newly opened The Canadian Petroleum Hall of Fame in Leduc, Alberta. He has spent decades researching, encouraging academics and media, giving memorable speeches, and sitting on countless boards and committees.

Since its inception in the late 1970s, he has been instrumental in developing Petrolia Discovery. For two decades, he had been donning his work boots and oil-splattered clothes to labour in Discovery's oil field. Although he's a busy man, he has endured countless questions at all hours in the writing of this book. The importance of setting on paper the knowledge he carries in his head was never doubted.

When the West Virginia University industrial archaelogists arrived at Fairbank Oil, they needed a "little background information."

Introduction

Oil is the second most abundant fluid in the world. Only water is more plentiful. It has been said that among all the legal businesses in the world, oil ranks as number two. This enormous industry that spans the globe has risen so far and so fast that it seems impossible it began in Oil Springs, Ontario, Canada. But it did.

Most history books today still credit Col. Edwin Drake for being the first to strike oil near Titusville, Pennsylvania in 1859. The work in Oil Springs, however, precedes this. Although this is well known within Lambton County, it is not known at all in most of Ontario or Canada. Painstaking research by Col. Bruce Harkness, the former Natural Gas Commissioner for the Department of Mines in Ontario and other academics, documents how the oil industry actually began in Canada, in Ontario and in Oil Springs.

John Henry Fairbank is part of that story. By 1890, he had grown to be Canada's single largest oil producer. No one in this country was pumping more oil out of the ground. "Oil magnate, banker, hardware dealer, wagon builder and sometime member of the Canadian parliament, in his own way he achieved a pre-eminence in the Petrolia area parallel to that of a Carnegie or an Eaton," wrote historian Edward Phelps in his thesis, *John Henry Fairbank of Petrolia* (1831 - 1914) A *Canadian Entrepreneur.*

J.H., as he was usually called, became such a leading light in Petrolia that his accomplishments there have overshadowed his oil field legacy in Oil Springs. The oil business he founded in Oil Springs back in 1861 has survived and even thrived right through to today. It has been passed down from father to son through three generations, making the Fairbanks the oldest petroleum producing family in the world.

It seems impossible that today's enormous oil industry began in Oil Springs, Ontario. But it did.

The Fairbanks have also been supplying crude oil to Imperial Oil longer than anyone. It was always thought to be probable, but until recently no written proof could be found. In a family trunk, a document was unearthed. This "memo of purchase" shows that J.H. was selling 4,700 barrels of crude to Imperial Oil in August, 1880 - one month before the famous company was officially incorporated and less than four months after Imperial Oil was first formed.

"The really important thing is that Fairbank Oil is using mid-19th century technology on a daily basis. There's nothing like that anywhere."

Today, more than 140 years after J.H. started drilling in Oil Springs, Charles Fairbank Oil Properties Ltd. has expanded to more than 600 acres with 350 working oil wells. Each year, this oil field renders about 24,000 barrels of oil with the help of a steady work crew and faithful maintenance. (In the Imperial system used in Canada, a barrel is 35 gallons; in the U.S. system it equals 42 gallons.)

Owned and operated by J.H.'s great-grandson, Charlie Fairbank, the oil field still uses the "jerker-line system" of pumping several wells together and sharing one power source. This primitive-looking system is quite ingenious and is the product of J.H. Fairbank's fertile mind.

Visiting industrial archaeologists, who study the early workings of industries and structures, are surprised to find this 19th century technology is not simply intact but it is used around the clock each day of the year to pump the wells.

In 1999, Dr. Emory Kemp and Dan Bonenberger, of West Virginia University's Institute for the History of Technology and Industrial Archaeology, led a team that spent several weeks at Fairbank Oil recording the oil equipment in archival detail.

When researching industrial systems of days long ago, these academics are often baffled by missing pieces of the puzzle. "When they came here it was a different story," says Charlie Fairbank. "It was as if they had been studying dinosaur fossils and stumbled upon a living, breathing dinosaur."

Fairbank Oil Properties is unique. "I believe that the modern petroleum industry can trace its origin to Fairbank Oil and the surrounding oil district," Dr. Kemp said. "The really important thing is that Fairbank Oil is using mid-19th century technology on a daily basis. There's nothing like that anywhere."

The equipment and technology at Fairbank Oil was once common throughout the Oil Heritage District.

Today, it's a preserved pocket of the past in a modern world. After the oil booms in both Petrolia and Oil Springs, the evidence began disappearing as the decades rolled by.

There are a number of sites on Fairbank Oil Properties to pique historians' interest. Among them are the wells of Charles and Henry Tripp which were dug by hand in the 1850s. Charles Tripp entered the pages of history when he developed asphalt and with it, he founded North America's first commercial oil business in 1854. Refining crude, to obtain oil, was not developed until 1858 by James Miller Williams.

Also on the property is the 1862 site of Canada's first oil gusher, the Hugh Nixon Shaw well; the Black and Matheson "Flowing Well" of 1862; one of the last three-pole derricks left standing in the Oil Heritage District; the site of the Fairbank Gusher, the first gas gusher of Lambton County; and the last of Imperial Oil's old receiving stations. "These sites," says Fairbank, "lay undisturbed, awaiting evaluation and recognition."

The significance of Fairbank Oil Properties could become much greater as the global oil supply begins to decline. Our galloping appetite for energy has sped us through the ages of horsepower, steam power and the age of coal. Now, we are approaching the end of the oil epoch. Although few are listening, experts are telling us that by 2010 the amount of oil extracted from the planet will begin to diminish. At some point in the future, Fairbank Oil Properties may represent not only the birth of the petroleum era but also its demise.

These historic oil sites "lay undisturbed, awaiting evaluation and recognition."

Already, academic interest in Fairbank Oil is growing. In the fall of 2000, about 70 members of the Society for Industrial Archaeology from many parts of United States toured Fairbank Oil and found it engrossing. More research was started in the summer of 2001 when Dr. Kemp returned to Fairbank Oil and conducted a six-week field school.

Parks Canada has been working closely with Dr. Kemp and groundwork is being laid for a national heritage designation. This would include various sites or areas of the Oil Heritage District in Lambton County. Dr. Kemp believes Charles Fairbank Oil Properties qualifies for a United Nations World Heritage Site designation and he is actively pursuing this cause.

The story of oil's early development had a cast of thousands. Men converged on Oil Springs to play both major and minor roles. This particular tale opens in 1861 when John Henry Fairbank arrives in Oil Springs and it winds through the decades right to the 21st Century. It's a thin wedge of time but a historically meaty one.

It's a thin wedge of time, but a historically meaty one.

One quick glance at the timeline at the back of the book gives an inkling of the monumental changes and events that completely reshaped our world. It chronicles many arrivals – kerosene light, the railways, the automobile, the roads, Canada's first pipelines and electricity. These leaps in technology occurred with a backdrop of wars, depressions and fires, as well as an ever changing cast of colourful characters. All of these had indelible consequences for Fairbank Oil in Oil Springs. Being a family business, the actions and accomplishments of family members also rippled through the decades and these are recorded here too.

From an age when horses provided transportation and whale oil gave light, oil has catapulted us to time of superhighways, private jets and cities that never sleep. There is quite a story here, but the full tale is yet to be written.

PART ONE
The Four Generations of Fairbanks

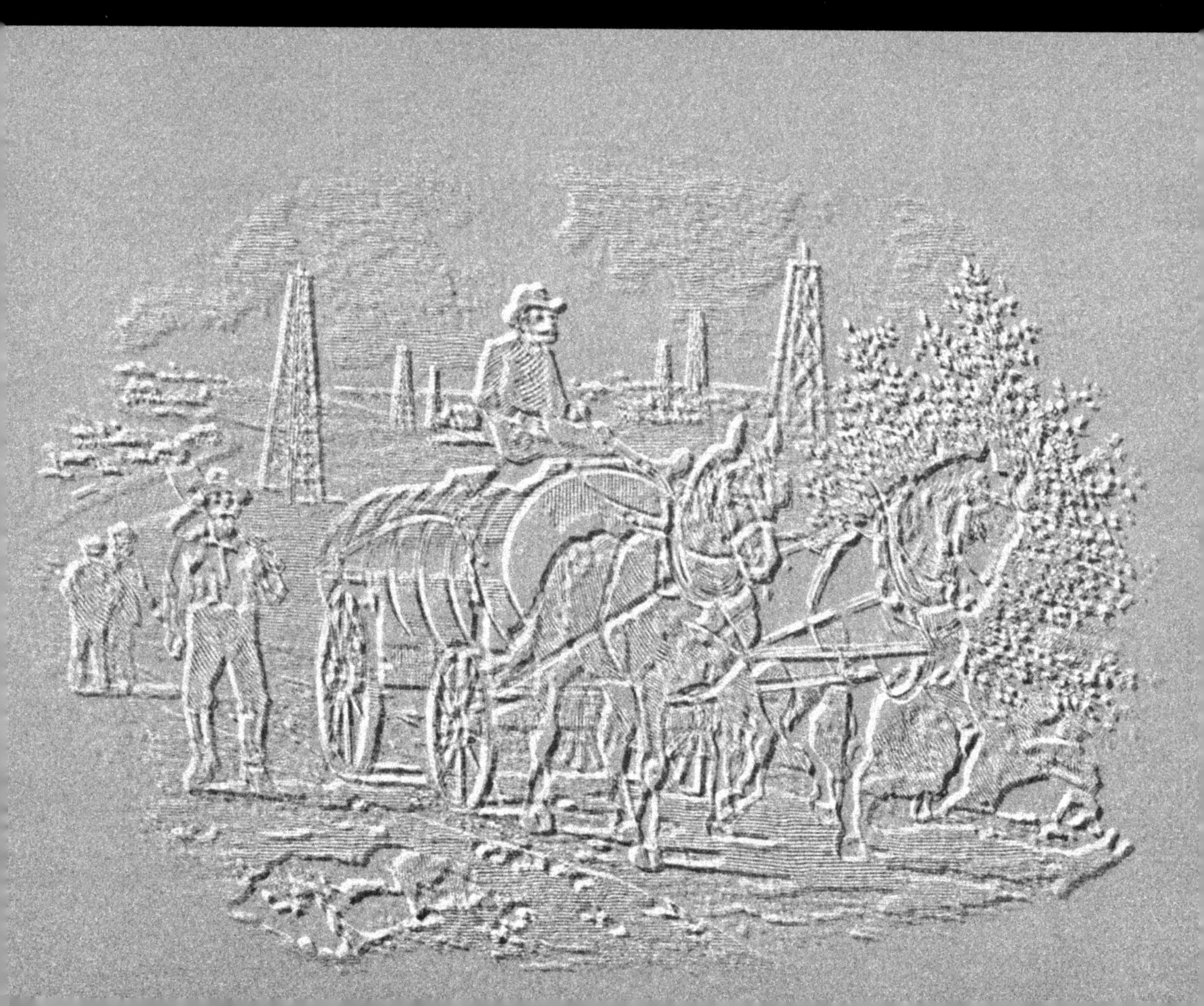

Chapter 1 – The Beginnings

Oil For Lighting Lamps

It began, as many great ventures do, with a simple quirk of fate. John Henry Fairbank was a 29-year old American with a young Canadian wife struggling on a farm in Niagara Falls, Ontario when a surveying job floated into his lap. Would Mr. Fairbank kindly survey a property in a place that had been newly named Oil Springs, 150 miles distant, for the wealthy Mrs. Julia Macklem?

She had just acquired the property from the cash-starved Charles Tripp, the man who first discovered the oil and founded North America's first commercial oil business. The survey would require a scant few months and the pay, though modest, would be money in hand. He had his wife, Edna, and two young sons to think about. Henry Addington Fairbank was nearly five years old and Charles Oliver was almost two.

When J.H. Fairbank said "yes" to Mrs. Macklem's job offer he unknowingly set foot on a path that would make him the largest oil producer in Canada, a politician, and the very wealthy head of a multi-faceted business empire. In time, the Fairbank family would earn the distinction of being Imperial Oil's longest-running supplier of crude oil. The earliest available record shows J.H. sold crude to Imperial Oil on August 12, 1880. Today, more than 120 years later, Fairbank Oil continues to ship crude to Imperial Oil.

J.H. Fairbank unknowingly set foot on a path that would make him the largest oil producer in Canada.

It was March 1861, not quite two years since the tiny village of Black Creek had been renamed Oil Springs. The assignment was to survey 100 acres of bush land and parcel it into 198 lots for the men flocking to the area to drill oil. News of the oil discoveries in Enniskillen Township was already rippling out far and wide. The year before, Leonard Baldwin Vaughn made headlines in *The Sarnia Observer* when oil from his well poured into Black Creek,

As these men sweated over their shovels digging for oil, they were seeking oil for lamps.

reportedly at a rate of 300 gallons each hour. Toronto's daily newspaper, *The Globe*, had already dispatched reporters to Oil Springs and the telegraph, which had arrived in the 1840s, disseminated the news quickly and economically.

When J. H. Fairbank arrived that spring, about 500 men in Oil Springs had started making their fortunes by striking oil and then refining it for lamp oil. A good many were from the United States where the four-year Civil War was about to begin. A few of these rugged adventurers had already tried their luck in the California gold rush in the 1850s.

The discovery of oil was so new that many of its uses were not yet realized. As these men sweated over their shovels digging for oil, they were seeking fuel for lamps. Everyone recognized the enormous potential for inexpensive oil that would provide homes and businesses with lamplight at night.

Until the 1850s in North America, most people lit tallow candles after the sun went down. Occasionally, explosive camphene distilled from turpentine was used in lamps and whale oil was quite common. Things were changing rapidly.

Across the Atlantic, gas produced from coal was being used for lighting and as early as 1807 the first gas lights were installed in London's Pall Mall. By 1852, it was introduced to the British House of Commons but would not be affordable to the British masses for decades. British whale oil companies tried to stop gas light claiming it was "dangerous, poisonous – or a defiance of Almighty God." British North America was early to embrace the wonder of gas lights too. Toronto first used them in December 1841; Hamilton had them by 1851 and London by 1854.

Shortly before this, in the late 1840s, a fantastic new lamp oil arrived on the scene in Halifax, Nova Scotia. Abraham Gesner of Nova Scotia, the first government geologist in the British Colony, had just created an oil made from coal. He called it "keroselain", derived from the Greek words for "wax oil" and soon afterwards it was renamed "kerosene". As well as using coal, Gesner distilled "Trinidad pitch" and "shepody pitch" and "asphaltum".

The coal oil was excellent for lamps and kerosene made a very good lubricant too, but what really mattered was the process he invented. First he heated coal to high temperatures, cooled it, then treated it with sulphuric acid and calcinated lime to remove the impurities that made it smoke and smell. Gesner simply called his process "distilling"; others would call it "refining" and say that Gesner invented the world's first oil refinery.

By 1854, he gained three U.S. patents for "Improvements in Kerosene burning fluids". He then set up a factory in New York State, called the North American Kerosene Gas Light Company, where he could access better raw materials. It wasn't long before crude oil superseded coal in Gesner's distilling process. Crude oil was cheaper and because it was a fluid, it was easier to distil.

Forever after, he has been celebrated as a founder of the modern petroleum industry. Gesner may have been the one bestowed with the credit but he realized that "the honour of discovering the process of refining oil is divided among independent workers in several countries," according to the unpublished 1940 manuscript, *Makers of Oil History 1850 to 1880* by Col. Robert Bruce Harkness, former Natural Gas Commissioner for the Department of Mines in Ontario. Geologists and historians have long recognized Col. Harkness as the leading authority on the early oil industry.

Imperial Oil has chosen to celebrate Gesner and its tribute at his Halifax grave says, "He gave the world a better light." The oil company had a direct tie to Gesner. His kerosene company was eventually bought by J.D. Rockefeller's Standard Oil. Standard Oil would later gobble up Canada's Imperial Oil.

During the 1850s, another stable oil was being imported from the U.S. – something called "burning fluid" which mixed camphene and alcohol. Lighting oil, once a luxury, was quickly seen as a necessity.

When James Miller Williams' "burning oil" from Oil Springs hit the market in 1858, the excitement was palpable. "A beautiful burning oil," is how *The Sarnia Observer* described it December 30, 1858. "Its illuminating properties are so great that an ordinary sized lamp giving a light equal to six or eight candles, can be kept burning at the rate of one-quarter cent per hour."

This was not the first account in *The Sarnia Observer*. On August 28, 1858, it had reported "an abundant supply of mineral oil which the owners of the land were taking steps for making available for the purpose of light, etc. by erecting works there on for purifying said oil and making it fit for use."

"A beautiful burning oil...giving a light equal to six or eight candles for one-quarter cent per hour," is how *The Sarnia Observer* described the oil found in Oil Springs.

***The Sarnia Observer* was writing about oil in Oil Springs in August, 1858. It was not until August, 1859 that Col. Edwin Drake discovered oil in Pennsylvania.**

These dates are important for they show that oil development happened in Canada before Col. Edwin Drake discovered oil in Titusville, Pennsylvania in August of 1859. Citing these newspaper articles Harkness wrote "...there can be no question that J.M Williams had established a petroleum industry as early as 1858 and most probably in 1857."

Writing in an Oil Springs centennial celebration brochure in 1958, Harkness went on to add that Williams must have been familiar with Tripp's process. "He (Williams) must have worked out every step through production, transmission, refining and selling a high grade illumination oil against the competition of the many manufactured 'burning fluids' and 'coal oils' on the market. Without question he was the first man to do this."

Williams' accomplishments were recognized at the International Exhibition in London, England in 1862. He received one medal for being the first to produce crude and a second medal for his refined oils. "This exhibition was open to all nations and should establish Williams priority without any question," wrote Harkness.

In Canada, if not in the U.S., Williams is hailed as "The Father of Refining", credited as being the first in the Dominion to produce crude oil and refine it. It was only by refining crude oil that it could be made into lamp oil. Williams went a step further than Charles and Henry Tripp, who had arrived in Oil Springs in 1850. The Tripp brothers found the oil of the gum beds and started making asphalt from it. The process they used allowed lighter fuels like naptha and kerosene to evaporate. By 1854, the Tripps finally got the Legislature approval to incorporate North America's first commercial oil business, The International Mining and Manufacturing Company.

Reportedly possessing better business acumen than the Tripps, Williams started buying up Enniskillen Township land very early. By 1856, he owned 600 acres and added another 200 the next year. When the Tripps' business failed, Williams took ownership of it and by 1858 he was making headlines with his burning fluid.

News of William's exciting work reached United States and came to the attention of a man named Colonel A.C. Ferris. He's the man known as "the long-awaited pioneer in introducing American petroleum to the kerosene manufacturers on a large scale" according to Allan Nevins in his well-respected two-volume biography entitled *John D. Rockefeller, The Heroic Age of American Enterprise*. Ferris is described as "a man of means and abounding energy" who, in 1857, happened to see a very effective tin lamp in Pittsburgh that used burning oil from the salt wells at Tarentum.

"He knew that the New York Kerosene Company was making a similar oil from coal. Without loss of time he flung himself into the field of supplying refiners with crude oil and marketing the product," Nevins wrote. "As the demand increased, the enterprising colonel began to search widely for supplies. He visited districts where crude oil was still gathered by the blanket-wringing method. When he heard that one J.M. Williams had opened an oil field in Ontario with a pick and a shovel, he hurried off to see him and bought his whole supply."

Almost a century after Ferris' trip to Ontario, the editors of the Canadian periodical *Oil-Gas World* examined this quote closely and decided it would not have been Williams' whole supply which Col. Ferris purchased, but his surplus. They reasoned that Williams was operating his refinery in Hamilton and selling kerosene as far west as Sarnia. (*Oil-Gas World* devoted its June, 1958, issue to celebrating 100 years of oil.)

Col. Ferris and what Nevins calls his "unremitting energy in popularizing the new illuminant" would have certainly caught the attention of J.D. Rockefeller. "Ferris did much to better the quality of lamps, while the New York Kerosene Company steadily improved the process of refining." Rockefeller would enter the refining business in 1862.

"When he heard that one J.M Williams had opened an oil field in Ontario with a pick and a shovel, he hurried off to see him and bought his whole supply."

It's quite likely that J.H. Fairbank knew of Williams' discoveries before he arrived in Oil Springs in 1861. Seven years earlier, a surveying job took him to Komoka, 50 miles north east of Oil Springs. He may also have heard that in 1859 Col. Drake had found oil in Pennsylvania.

Certainly by the spring of 1861, J.H. was well aware of the significance of the inexpensive burning oil. Whale oil was commonly used when J.H. arrived in Oil Springs and a gallon of it likely represented a man's daily wage. He later told the 1890 *Royal Commission on the Mineral Resources of Ontario and Measures For Their Development* that whale oil was then selling for $1.25 a gallon.

"By 1850, whale oil, considered by many to be the finest illuminant and also widely used as a lubricant, rose dramatically in price as the great mammals became rare and whalers had to search further to fill the holds of their ships."

Whale oil had not always been so expensive. "By 1850, whale oil, considered by many to be the finest illuminant and also widely used as a lubricant, rose dramatically in price as the great mammals became rare and whalers had to search further to fill the holds of their ships," wrote Thomas Carpenter in *Inventors – Profiles in Canadian Genius*. "The extravagant prices encouraged a widespread effort to find a substitute, and it was common knowledge that the discoverer of an alternative illuminant would become very wealthy indeed."

J.H. also astutely realized that oil would have a major role to play in industrialization, which was still in its infancy but destined to grow exponentially. In both Canada West and Canada East there were foundries and paper mills by 1850. As well, steam engines and farm machinery were in production and they would be key markets for lubricating oil.

Steam-powered railways had just begun to revolutionize transportation. Although Canada's first railway opened in 1836, it wasn't until the 1850s and '60s that substantial distance of track was laid. The Great Western Railway connected London and Sarnia by 1858 and within two years, the line was continuous to Montreal.

A few short months before J.H. arrived, another American living in Oil Springs, Leonard Baldwin Vaughn, had already met considerable success. On October 8, 1860 one of Vaughn's wells produced a vast amount of oil and an equal amount of exhilaration. This was the first of the rare "flowing wells", the ones which needed no pump.

In February of 1861, census enumerator John Smith noted in his report that in 1854, there had been only 54 people on the assessment role for the whole 86,000 acres of land that made up Enniskillen Township. "But," he wrote, "The oil mania has brought the greatest influx of population to this township. It commenced in April last and has tended to increase the inhabitants and to put a fictitious value on land, many parcels of land that a few years ago were thought hardly worth the taxes are now held at high prices."

Clearly baffled by the whole oil business, he went on to write, "The problem to be solved at present is does this substance abound in the crevices of the rock generally or is it confined to those spots where it has showed itself spontaneously? For the purpose of solving this problem, wells are being sunk in various places but hitherto they have not been successful in any place at a distance from where the oil appeared naturally." (see Asides, pg. 181.)

In the census of 1861, those enumerated were required to give their job titles. No one is listed as an oil producer; instead the job was described as "miner".

"Hundreds of people are coming in every week from all parts of America, most of them from Ohio and Pennsylvania," reported Toronto's *Globe* in 1861. "There is no doubt of it – a source of untold wealth is in our midst."

The eyes and ears of J.H. were filled with this day after day while he surveyed, and within a few short months, he realized resistance was futile. The possibilities were just too irresistible and he too, was caught up in the cyclone of excitement.

Seduced by the Lure of Oil

Despite J.H. Fairbank's unbridled enthusiasm for oil, his wife, Edna, was clearly unmoved by this oil business. To her chagrin, J.H. borrowed heavily and threw in their meagre savings to finance the digging of his first well. Records show that on July 15, 1861, he leased a half-acre lot in Oil Springs for $300 and broke ground on his very first well. He christened it "Old Fairbank".

Pumping the oil up from the belly of the earth required enormous exertion. In Oil Springs, it meant digging down about 50 feet into heavy clay with a shovel, loading the clay into buckets, hauling the buckets up to the surface with a pulley, unloading them into a wagon and having a horse cart the clay away. As the work progressed deeper, they "cribbed" the hole with a liner of logs. This cribbing made the walls stable and kept out debris. Then, the well had to be fitted with a long, complicated series of pipes, casings and finally, a pump.

The pump needed to be powered by one of two methods. A man could build a "spring pole" out of the black ash trees which gave a good bounce. This device could pump the oil when a man repeatedly jumped or "tramped" on the treadle. (Note, the spring pole worked as a pump but it could also be modified so that it chiselled into the rock.) The alternate method of pumping oil was outfitting each well with a steam engine, but this was expensive. All this work was no guarantee of oil. It could be a dry hole producing no oil at all.

Most property lots measured about a half-acre. The oil pioneers usually dug

J.H. Fairbank began his illustrious career in 1861 by leasing a half-acre lot in Oil Springs for $300 and breaking ground on his very first well, "Old Fairbank".

their first well in the centre of it and if it produced oil, they would go on to dig four more wells, one in each corner of the lot.

Getting the oil pumped to the surface was only part of the battle. J.H. and other oil producers discovered it was equally exhausting to haul the barrels through 13 miles of swamp to the railhead in Wyoming on the Sarnia-London line. A horse-drawn wagon could only take only one or two barrels at a time in spring or summer. In winter, a team of horses could haul 14 to 16 barrels on a sled through the snow.

"Wagons pushed aside into the bush or still sticking in the mud...tell where attempts to reach Wyoming or Oil Springs were abandoned in despair."

In the 1800s, the geography was so intimidating that parts of Lambton County and a good deal of Bruce County were among the last areas of Ontario to be settled. Hardly anyone came to places like Enniskillen Township in the interior of Lambton County because they were considered uninhabitable by most.

"To be compelled either to walk or ride twelve and a half miles between (the railhead at) Wyoming and Oil Springs is a dreadful calamity," reported *The Globe* in 1861. "The number of mud holes is something wonderful. Not little paltry affairs, giving the vehicle a slight twist and a jerk but 'big things', large enough for a horse to swim in. But swimming is prevented by the stiff clay mixed in with a promising, firm-looking crust on the top, seemingly capable of bearing the weight of the greatest man in this magnificent province; but deceptive, gifted as it would appear with the powers of suction, burying horses up to their shoulders, and retaining wagons firmly in its grasp...Wagons pushed aside into the bush or still sticking in the mud, and piles of lumber on the road, tell where attempts to reach Wyoming or Oil Springs were abandoned in despair."

Another report, from a *Toronto Leader* reporter in 1862, had been more encouraging: "I came across a well every few rods. There in the wild woods were hundreds of men, all quiet, intent upon their work. There was no talking but the tramp, tramp, tramp of the foot (on the spring pole). Click, click, click, the sharp sound of the drill as the steel bit its way into the rock.

"In a few places in the woods, we encountered buildings of more pretentious appearance from the majority. Here were wells worked by steam engines. How they ever got in, I know not. But here they are, working day and night," the reporter marvelled.

Early letters show Fairbank largely attacked the challenges with a certain degree of zeal while Edna pleaded with him to come to his senses and return home.

Transportation remained a harrowing problem but improved somewhat when a "plank" road made of logs was completed from Oil Springs to the railhead in Wyoming in May 1862. "There followed a great rush of traffic on the plank road; it is reported that there were 400 teams hauling oil, and the loads averaged 16 barrels with 20 barrels as an outside load," wrote Col. Harkness. "Under such traffic conditions the confusion along the road and especially at Wyoming can scarcely be imagined."

Within a year, the road was so badly worn out that it needed $6,000 in repair, but there were no funds for rebuilding it. This road would be the precursor for what is now called The Oil Heritage Road, formerly Hwy 21. It was not paved until 1934; the job was a work project during the Depression.

Two years after this first plank road, another was built to connect Oil Springs to Sarnia. It too was in extremely bad shape after one year of heavy use. Today, this road is still known as The Plank Road.

The landscape of Enniskillen Township today bears little resemblance to what the early oil men saw. It doesn't even resemble what Bill Abraham saw when he became the Ministry of Agriculture's representative for Lambton County in 1958. Beginning as far back as the late 1860s and continuing to this day, extensive networks of drains and tiles have been laid underground to extract enormous quantities of water. Under the surface of Enniskillen and Brooke Townships, there are more than 500 miles of drains. A swamp has been transformed into arable land.

The land here is called Heavy Brookston Clay. "Lambton County is pretty much one solid mass of it," says Abraham, who dealt with drainage daily throughout his 25-year career with the ministry. "It has definitely been one of the wettest areas in the province. The best example of what it does was to be found at the Brigden Fairgrounds. In the late 1950s there, you'd have a bit of mud on your boots in the morning and you'd be up to your crotch in it by late afternoon." One of his predecessors had a wretched time, for his Model-T Ford was forever getting stuck.

In the 1860s, horses were essential for transporting oil and hence, very

The land here is called Heavy Brookston Clay. "...you'd have a bit of mud on your boots in the morning and you'd be up to your crotch in it by late afternoon."

numerous. In addition to the 400 teams hauling oil to Wyoming, horses were needed for the stagecoaches that ran four times daily between Oil Springs and Sarnia. Within Oil Springs, horse-drawn wagons ferried people along the main street at five-minute intervals at busy times of the day and less frequently in the evening.

"I timed the filling of these barrels and found that in one minute and 45 second, each barrel was filled."

Riding the Oil Springs Boom

Oil Springs in the early 1860s had a charged atmosphere; the highly contagious oil fever ran high. In these exhilarating days, not every man kept a firm grip on logic, some became slightly crazed. A man named Hugh Nixon Shaw came to be considered at least half-crazed. He appeared sane enough when he arrived and even helped three other men design the layout of the village.

He earned the moniker "that Crazy Shaw" by steadily jumping on the treadle of his spring pole for six months straight. He had already spent his last dime, yet he kept his iron bit chiselling deeper than any rational man thought necessary. It was thought to be sheer folly that he chiselled right into the second horizon of porous limestone 158 feet below bedrock.

Shaw hit pay dirt on January 16, 1862. A gusher exploded and oil shot 20 feet into the air. It was a sight never seen before in The Dominion, a first. T*he Globe* of January 28, 1862 carried a report on the event. "On the 16th of this month a Mr. Shaw, lately of Port Huron, Michigan, a daguerrean artist and formerly of Kingston, Canada West, struck oil as it is termed...50 feet through clay, 158 feet in the rock... Within 15 minutes of the last drill of the chisel, the oil was overflowing the surface of the earth, the well being entirely filled...Four 120 barrel tanks were filled and many barrels...I timed the filling of these barrels and found that in one minute and 45 seconds each barrel was filled...From the time that well was brought into partial subjection up to this time, upwards of 2,000 barrels have been taken away and more than this quantity has been lost...Owing to the want of barrels, only some 650 barrels are taken from the well each day." (The barrel, a measurement still used today, is equivalent to 158 litres, 35 Imperial gallons and 42 U.S. gallons.)

Americans who witnessed gushers in Titusville, Pennsylvania, had managed to cap the well by stuffing it with a "packer" made from calfskin and flax.

It's not clear how many days passed between the gusher being struck and the well being capped.

Shaw was no longer "crazy"; he was instantly rich. It was the others who went mad after his gusher. It was a classic rags-to-riches story, likely Canada's first. The very morning of the gusher, Shaw had been refused credit for a humble pair of boots.

In March of 1862, J.H. wrote Edna that Shaw had refused an astonishing offer of $10,000 for his well. Oil fever ran rampant and the population of Oil Springs swelled. Before the year was out, there would be 1,000 wells in Oil Springs producing 12,000 barrels of oil daily.

Although Shaw has always been credited as being the first to chisel below bedrock, Col. Harkness dug up evidence that it was really Williams who did it first. He cites a detailed report in *The Globe* of September 6, 1861 and concludes, "His facts show that Williams drilled the first producing well in the rock sometime in 1859."

Shaw's was no ordinary oil well. It was the one of the 33 "great flowing rock wells" to be discovered and the most prolific. The Shaw well only lasted ten months, but in that time, it produced an impressive 35,000 barrels of oil.

These "flowing wells" tapped into huge reservoirs of oil and they flowed freely, not ever needing a pump. It has been said they created actual oil floods. Anyone standing within the 40 to 50 acres surrounding them would be soaked to their calves or even hips according to *Belden's Historical Atlas of The County of Lambton*, 1880. During their first few months, the flowing wells reportedly blackened the ground with one to three feet of oil.

This paints an incredible image. In fact, it's so incredible that it is now considered rather questionable. The news in Oil Springs was fantastic enough, yet it now seems that the writers of the *Belden Atlas* felt compelled to embellish the facts.

Another report of the scene in 1862 came from Alexander Winchell, a leading American geologist of his day. His book, *Sketches of Creation*, published in 1872, listed 33 "flowing wells" in Oil Springs. He gave fairly precise locations of them and their daily flow capacity in barrels. One J.H. Fairbank well is listed as producing 500 barrels a day while the wells of 18 other men are recorded with much higher flows.

"There was no use for oil at that time," Winchell wrote. "The price had

Shaw was no longer "crazy"; he was instantly rich. It was the others who went mad after his gusher.

"A house without hope" is what Edna called the plans for J.H. to build a shanty in Oil Springs .

fallen to ten cents per barrel...Some of these wells flowed three hundred and six hundred barrels per day...Three flowed severally (sic) six thousand barrels per day; and the Black & Mathewson (sic) well flowed seven thousand five hundred barrels per day! Three years later that oil would have brought ten dollars per barrel in gold. Now its escape was the mere pastime of full-grown boys. It floated on the water Black Creek to the depth of six inches and formed a film upon the surface of Lake Erie." (Winchell likely meant the St. Clair River, not Lake Erie.)

Winchell's calculations that 5 million barrels of oil from Oil Springs flowed unchecked are now highly doubted, yet the authoritative *Geological Survey of Canada of* 1890 seemed to have no qualms about his figures and printed them as assured facts.

Within months of Shaw's gusher, Fairbank saw he was far too engrossed in oil to even consider returning to his former life in Niagara Falls. He stopped renting his living quarters and built a log shanty measuring 12 by 16 feet, a size that's smaller than many modern living rooms. This meant he no longer viewed himself as a temporary visitor; he was declaring himself a resident.

Edna was forced to face the fact that oil was not a passing fancy for her husband but a major obsession. And although he wanted her to join him, she balked at leaving the comparatively civilized life of Niagara Falls for a more rugged existence in the evil sulphurous smelling Oil Springs. "A house without hope" is what she called the plans for the shanty in the spring of 1862. The correspondence between the couple does not indicate Edna ever visited Oil Springs even though J.H. would live in the shanty until 1865. Instead, J.H. returned to Niagara Falls a few times each year.

In October of 1862, she wrote to J.H. of how she worried when he is "up in that hole" particularly when engaged in "that dangerous operation of digging for oil." And it was perilous. Men died in explosions, they died inhaling fumes and within the year, Shaw would drown in his well.

At the time of this particular letter, Edna had just learned that the drinking water was making J.H. sick and urged him to come home until the plank road to Sarnia was built, "thereby avoiding that miserable swamp during the wet, cold season."

Oil Springs was hardly an inviting

place at any time of year. She no doubt well remembered his description of the summer. In July 1861, J.H. had apologized to Edna for not writing earlier: "The reason was that it was so awful hot here that it was all we could do to live, and I fear that I shall not accomplish much tonight as I feel somewhat tired after my day's work and weak from loss of blood sucked out by some fourteen million mosquitoes."

Given the decidedly difficult living conditions, few men were bringing their families. There was a school in Oil Springs, a primitive building built of logs that had been floated down from the village of Florence, and the school was erected in 1854, a few years before the oil boom. A total of 13 pupils were enrolled in 1857. By 1865, Oil Springs was booming and there were more children than the tiny school could hold. The school was declared too small and the building was then rented to "Old Ellen", the housekeeper of a Dr. Macklin. Despite this evidence of at least some family life in Oil Springs, Edna would not be moved, literally or figuratively.

It should be noted that even when she was in Niagara Falls she wrote frequently of her frail health and possibly the hardships of Oil Springs would have been more than she could bear. Throughout the decades that followed, the couple's letters showed she was often sick or unwell with unspecified ailments. She would come to know considerable grief in her lifetime and this may well have attributed to some of her later physical problems. Edna would often take "hydropathic cures" or be recuperating for months at a time in England, Ireland, Rochester, Cleveland and eventually California.

The Edna of the 1860s was also averse to the financial risk that Oil Springs represented. In her letters to J.H., she argues they need to hang on to the family farm because the oil business was far too much of a gamble.

Some might assume that Oil Springs had a rollicking, testosterone-charged atmosphere, like the Klondike gold rush decades later, and this might be unappealing to women like Edna with young children. But this was not the case. "Although liquor is sold at most of the houses, there is scarcely any drunkenness," reported *The Globe* in 1861. "Those who have been there since the commencement bear witness to the fact the community is a very quiet one. No rows have taken place; knifings and shootings being entirely unknown."

"...I feel somewhat tired after my day's work and weak from the loss of blood sucked out by some fourteen million mosquitoes."

The technology was new, primitive and extremely risky - the stills were prone to exploding or bursting into flame.

J.H.'s mother, Mary Oliver Fairbank, was 75 years old in 1862 and had been a widow for 20 years, yet she had the pluck to pack up her things in Rouse's Point, New York, and move to this primitive place to keep house. John Henry Fairbank had been born in Rouse's Point, an only child. Mary Oliver had been his father's second wife; his first wife dying when giving birth to their first child. J.H. was named after his father's brother, John Henry Fairbanks, a fur trader "known to every Chippewa Indian in Minnesota...master of the Indian languages ...a man of high moral worth, strictly temperate in his habits, charitable to a fault, and noted for his tender affection for his children."

In addition to arranging that J.H.'s mother would join him in Oil Springs, it was decided that his 4-year old son, Charles, would live with them too. Henry, two years older than Charles, would stay with his mother, Edna, in Niagara Falls. It is not known why this particular arrangement was made.

It's testament to J.H.'s faith in his oil wells that he chose to build his shanty and commit to Oil Springs even though he was desperately broke. His property was valued at $2,000 but his equity in it was a mere one-quarter. Also, he badly needed $300 to meet his payments on land and equipment. Fortunately, his father-in-law, Hermanus Crysler, was ready to join the great gamble and he mortgaged his Niagara Falls real estate to provide much needed cash. Mary Fairbank was also loaning money to him.

During the summer of 1862, Fairbank spent three months building a refinery. He wrote that he was using the "Hugh Shaw Patent Still" that fastened two large kettles together to form a sphere. The technology was new, primitive and extremely risky - the stills were prone to exploding or bursting into flame. It's not clear if this still had a true patent. No records or diagrams of it can be found.

After a particularly frustrating day in October, he poured his angst into his journal. "...About as miserable a day as I ever put in, run till dark and quite fully realized that I won't run a damned leaky old kettle that acts as if it would 'go up' at any minute for love nor money - don't want to become as nervous as an old maid, and feel like a coward all the time." He continued however, to run the refinery for about two more years, refining his own oil and charging others eight cents a gallon for the service.

While the years of 1862 and 1863 were mostly hard scrabble for J.H., they were good years for Oil Springs. *The Oil Springs Chronicle* issued its first newspaper on April 23, 1862. The next year, the village boasted its very own telegraph service, a new school, and it erected two churches.

It was only after J.H. endured a litany of struggles during his first two and a half years in Oil Springs that finally the elements of luck, hard work and good judgement had mixed together to a point where he could actually get a glimpse of success. Old Fairbank proved to be an excellent well and in November 1863, Fairbank gleefully wrote in his journal that "good boy" had produced 45 barrels of oil in 24 hours. "Net profit of day $150, a big day's work, the biggest ever made by me or probably that I shall ever make." This sum likely represented three or four month's wages for a hired hand at that time.

Success may have glimmered ahead but he was drowning in debt. Four times in 1863, he was sued for failing to meet payments on his oil field. James Miller Williams, the man who is hailed as the father of the oil industry, trotted J.H. off to court for a debt of $281.30 and won, leaving a disappointed Fairbank with the court costs of $52.81 as well.

When George Taylor sued him for a $350 debt, J.H. concocted a clever ruse to keep him waiting. Fairbank urged his own mother to sue him for $400 knowing that this higher amount would take precedence in the Court of Common Pleas.

The Sheriff of Lambton County presided over the auction of his goods and was convinced it was a true sale. In reality, Fairbank's friends bought the auctioned goods with the understanding Mary Fairbank would promptly buy them back. He later paid back Taylor. The fourth time, he was sued for $300 and was able to stall for time until he had the cash in hand.

Keeping his head above water in 1863 was a major feat. They were financially precarious times even though fortunes were sometimes made, lost and regained within days. Fairbank could have gone under at any time and like so many of his day, remain an unknown. Oil prices fluctuated wildly but his oil was pumping and Oil Springs was bustling. But in the spring of that year, the steady gushers, those "great flowing wells", started spouting salt water instead of oil.

Time would soon prove the boom had crested but not died. Oil from less

They were financially precarious times even though fortunes were sometimes made, lost and regained within days.

spectacular wells was still so abundant in Oil Springs that a well producing only five or six barrels a day was considered insignificant and duly abandoned. (This contrasts sharply with today. In mid-May, 2003, a barrel of oil was priced at $28.50 and production of five barrels daily for one year would be valued at $78,018.75 Canadian.)

Worried oil producers had their doubts confirmed by a report in *The Oil Springs Chronicle* on Oct. 23, 1863. "The local boom continues to level off. The wells flow salt water…the Jury and Clark and Evoy well, the most valuable for a long time, failed a day or two previous to the above date. The field is diminishing…" By the end of 1863, the glut of oil sent prices tumbling to 75 cents a barrel.

Many producers clung ot the hope that this was just a dip in the phenomenal financial roller coaster ride of oil. Those who didn't crash need to keep their wits about them. They also needed luck and a constitution that would handle the stomach-churning ride.

Like gamblers hanging onto their cards even when the chips are down, many producers clung to the hope that this was just a dip in the phenomenal financial roller coaster ride of oil. Those who didn't crash needed to keep their wits about them. They also needed luck and a constitution that would handle the stomach-churning ride.

Between 1862 and 1865 the price for a barrel of crude would plunge to 10 cents and soar to $10. "The work is very hard," J.H. later told the Royal Commission, "and requires a strong frame and a clear head."

Inventing on the Run

The oil business was so new and the problems so numerous that the men in Oil Springs were quickly inventing tools, technology and techniques as they went. It was engineering-on-the-run dreamed up by men who had abandoned farms, trades and businesses to try their luck at oil. There really wasn't a standard way to go about drilling oil and to a certain extent the men cobbled together their own techniques. Some were brilliant.

One problem they faced was the high expense of operating a steam engine for each well. Fairbank devised a system hooking 20 or more wells together with long wooden poles so they could share the power of one steam engine. The poles ran parallel to the ground and about a foot above it. Because it "jerked" as it went back and forth it became known as the "jerker system".

Later in 1890, Fairbank would tell the Royal Commission, "I remember the time the first jerker was put into operation. It was not patented, and I do not know that it could be. I had a well too hard to work by man power; I hadn't an engine, but there was engine power within reach and I applied the present jerker system. I think that was in 1863. The majority of the wells were then worked by man power with a spring pole.

"The jerker is universal now, and it would be impossible to work upon the old system. It was first used with a horizontal 'walking beam' that was afterwards improved by using the wheel, with which there is a great deal less friction. I think Mr. Reynolds introduced the wheel. With one engine they now work from a half a dozen to 80 or 90 wells, with one boiler but often two engines."

The impact of the jerker system is underscored in the 1917 book *Petroleum In Canada*. "The introduction of this system reduced the cost of pumping very materially," wrote author Victor Ross. "And the example set by Petrolia (sic) was quickly followed all over the United States."

Although it appears J.H. was the first to invent the jerker line, there is evidence of a similar system, the Strangenkunst, being used for mining in Germany as early as the 17th century. Dr. Emory Kemp, director of the Institute for the History of Technology and Industrial Archaeology at West Virginia University studied both systems. "I would have to say that J.H. Fairbank's jerker line was a case of independent invention," said Dr. Kemp. There is no evidence that the knowledge of this technology was able to cross the Atlantic Ocean at that time.

J.H. Fairbank devised "the jerker line system" so that one steam engine could power 20 wells at once.

Escaping the Oil Springs Bust

By 1864, Fairbank's financial picture had brightened considerably. Oil that had averaged at $3.50 a barrel jumped to $5 per barrel and his oil investments started bearing fruit. The amount of oil extracted daily may have been exaggerated somewhat in Winchell's reporting of 1862, but he did give an accurate picture of which producers were most successful and where they stood in terms of production. He showed that the Fairbank holdings were good but not nearly as terrific as many others in Oil Springs. Most of the other names have disappeared, their stories have gone untold, their lives unremembered.

As prices climbed, more outsiders were willing to jump in and pay premium prices for oil properties. A Chicago firm was even building a 108-

bedroom hotel on the main street of Oil Springs and it was touted to become the largest wooden building in the province. It was three storeys high, built in two wings, with a large verandah across the front, according to Arthur B. Johnston in a 1938 writing called *Recollections of Oil Drilling at Oil Springs Ontario.* (The location is not clear but it seems it was built on the north side of the main street and towards the east end of the village's business area).

Nine other hotels were doing a flourishing business in Oil Springs at the time and there were enough people to keep the cash registers busy at 12 large general stores states *Belden's Historical Atlas.* To all appearances, business was booming.

In the height of these heady days, when the Oil Springs population soared to 4,000, J.H. noted his oil production was dropping. Unlike so many around him, he silently accepted the fact that the Oil Springs boom would not continue. Quietly, he started selling several of his oil holdings in Oil Springs. He was like the thinking person who heads out early at the height of a great party, knowing bad weather is on the way. He already had his eye on oil lands in Petrolia.

His timing was impeccable.

His timing was impeccable. In 1865, he sold Old Fairbank at a very handsome profit. He had bought it for $1,000, extracted tidy profits for four years, then after talking to a man on a stage ride to Sarnia, he managed to sell it for $6,000. The payment was in gold and Fairbank told his son Charles that he carried the gold in a bag and happily marched into the Bank of British North America.

"The Old Fairbank in Oil Springs was about the only well pumping at the time," he told his son, Charles. "It had been pumping about 10 barrels a day and I kept it running very smoothly...Some parties from Little Falls, N.Y. bought it with other properties. They put in about $100,000, out of which they never got a dollar. They tried to improve my well by putting in a blower. They collapsed the conductor and spoiled the well."

Chapter 2 – Setting Hopes on Petrolia

J.H. knew exactly when to sell in Oil Springs but more importantly, he was purchasing land in Petrolia before most others. He was buying when prices were low and by 1865 he had already established his hardware store in Petrolia. Initially, it sold groceries and liquor and was built slightly west of Bear Creek. (See Asides, pg. 161.) It proved to be an excellent location because the main business section was first built at the bottom of Petrolia's east end hill. The upstairs of the store was known as Fairbank Hall and meetings of every sort were held there. It also served as the town hall.

In November 1865, he bought land for building a good-sized wood frame house west of Bear Creek and swiftly snapped up 50 acres of land east of his house property, subdivided it, sold it and pocketed $28,000 profit within months.

During 1865, his wife, Edna, was taking a health treatment for several months in Rochester, N.Y. and by November she was addressing her letters to J.H. in Petrolia. It appears he likely moved into the new house by late December that year.

While the house was being built, there was a feeling of unease throughout the area. The Fenians, the Irish-Americans seeking independence for Ireland, seemed poised to strike blows against Britain by raiding or "liberating" its colony. In the fall of 1865, the government of British North America ordered nine companies of the Active Militia (totalling about 500 men) to several posts along the U.S. border. Among the nine companies was the St. Clair Borderers, posted in Mooretown, along the St. Clair River.

When no attack came, the militia from the city of London was welcomed home with parades in April, 1866 but the unease lingered. The Fenians did attack, and they did it April, but it was

In November 1865, J.H. Fairbank bought land to build a good-sized wood frame house west of Bear Creek.

in New Brunswick and their force dissolved after the arrest of their leader.

Despite his rising fortune, it was not a grand house that he built in Petrolia, though it was a substantial step up from his Oil Springs shanty. Built of board and batten, it became the first house built in Petrolia west of Bear Creek and it stood at what is now the northeast intersection of Petrolia Line and Tank Street. (This original board and batten house still stands today. It was moved to the north of the Fairbank mansion to be a carriage house. A second wooden house would be built on the site of the original house and used by the family before the mansion was completed in 1891.)

By 1865, J.H. Fairbank had already established his hardware store in Petrolia.

There's no question, J.H. was early on the scene in Petrolia. In the spring of 1866, there were only four frame buildings west of Bear Creek; his and those belonging to men named Wheelwright, Col. Thompson and Bennett.

The Fairbank family was finally reunited on May 1, 1866 when Edna and their son Henry, now almost 9, moved into the Petrolia house to join J.H., their son Charles, and J.H.'s mother, Mary. They had just learned of Edna's new pregnancy and were settling into their new Petrolia home when fear struck.

On June 1, Fenians crossed the Niagara frontier, defeated the Canadian militia at Ridgeway and retreated. This would have felt alarmingly close; Edna's extended family was still in Niagara Falls. The following week, Fenians raided Quebec, staying only two days. A war between United States and Britain seemed imminent.

Panic ensued. Bracing for further attacks, the City of London had 300 men sworn in as the Home Guard on June 7 and 8, "armed with government rifles and being divided into companies, patrolled the streets of the city."

In Oil Springs, reaction was swift. The newly built 108-room hotel was abandoned overnight. "On the very day the plasters finished work upon the interior, the Fenian raid occurred, the American exodus took place at once and the new hotel was never even swept out," according to *Belden's Historical Atlas of* 1880. "Part of it has been pulled down to make use of the material for other purposes, and a part still stands, the home of bats, and rats and owls, instead of speculative travellers and live oil operators."

When performing a routine reality check on the "overnight" exodus from Oil Springs, Harkness drily noted there were other factors at play as well. "The plank road to Wyoming was in a deplorable condition, the operators were pumping over 90 per cent water and expenses from shutdowns were mounting. Such operations could only be carried on at a loss."

Undoubtedly, these were uncertain times, particularly for the leading oil producers who were largely American. No one knew what to expect next. The sense of unease and alarm wouldn't abate until 1867 when the uprising in Ireland collapsed.

The response to the Fenian threat, in retrospect, was wildly out of proportion to reality. "June 1, 1866, the fool Fenian raid occurred," wrote J.H. in the 1908 Old Boys Reunion souvenir book. "It was a cooling bath for Petrolia."

Although the first months in their Petrolia home were tense for the Fairbanks, everything brightened on November 23, 1866. This was a pivotal date, a date that would eventually spur the development of oil throughout the world. It's the date Captain B. King, of St. Catharines, struck Petrolia's first gusher. The location was found with the help of a well-known oil diviner, Mr. Kelsey of Buffalo, New York. (see Asides, pg. 177.) Once capped, the King well was producing a magnificent 265 barrels each day. The news hit like the crack of a starter's pistol and the Petrolia boom was off and running.

This gusher's impact would be far greater than Shaw's in Oil Springs. This impact would prove to be global. It was the King well discovery that triggered the Petrolia boom. And unlike the short-lived Oil Springs boom, this one spanned four glorious decades. Petrolia became the oil capital of Canada and boasted the highest per-capita income in the country. Until the turn of the century, this oil field provided almost 80 per cent of Canada's oil needs.

King's oil field was a pivotal moment in history because it changed the way energy was produced around the world. It gave birth to the "Petrolia foreign drillers" who took their knowledge of oil technology to far flung corners of the world and taught others how to open and run an oil field. Records show there were more than 500 of them and they went to more than 42 countries - among them were Egypt, Arabia, Russia, Persia, Poland, Peru, Venezuela and Borneo. This was remarkably exotic in an age when most people travelled no further than 30 miles in their lifetime.

"It is owing to that (the best quality

On November 23, 1866, Captain B. King struck Petrolia's first gusher with the help of a well-known oil diviner. The impact would be global.

"The return of a Petrolia driller from Persia after a journey of 13,000 miles by land or sea, the departure of a Petrolia driller for Burma or Australia, causes no excitement."

of equipment) and the skill of our drillers that Petrolia men are in demand all over the world," J.H. later explained to the Royal Commission in 1890. "We have drillers now in Germany, Austria, India, Burma, Mexico and Australia...I think more than a dozen have gone away within a month. The cause of the demand is that they have superior tools (designed and made in Petrolia) and possess superior intelligence. Our men have become great experts at it. By handling the pole they can tell what is going on down below 1,000 feet as well as if they were there."

Decades later, writer Victor Lauriston stumbled upon these facts found them incredible. He was hard-pressed to conceal his amazement in a story he wrote for *Maclean's Magazine* published May 1, 1924. Under the headline, The Town of World Travellers, Lauriston wrote: "Which is the most widely known community in Canada? Known, I mean, at least by name, to the greatest number of individuals, to the most varieties of humanity, and to the most widely scattered lands of the earth? If you were to ask me that question – if you were to bid me show you the favoured community - I would take you – Not to Ottawa, the capital; not to Quebec with its historic associations...nor to Montreal, beloved of the thirsty Yankee, nor to Toronto, Mecca of the Canadian intellectual but to a community of less than 5,000 people, in an isolated corner of southwestern Ontario...(to) Petrolia, the Canadian oil metropolis, and the headquarters of the Canadian oil drilling experts, who, during the past forty years, have pioneered for oil in all corners of the world.

"To Petrolia, travel is mere commonplace; as much of the everyday texture of life as eating and sleeping. The return of a Petrolia driller from Persia after a journey of 13,000 miles by land and sea, the departure of a Petrolia driller for Burma or Australia or the Argentine, causes no excitement. World travellers come and go every day – in Petrolia." Thinking that the readers may be doubtful, Lauriston felt compelled to add, "These statements may seem far fetched. Yet they are literally and indubitably true."

It would seem that the discovery of the King well, which spawned all these developments, should have plenty of documentation. But this has not been the case. There is even confusion about his name; he is sometimes referred to as Bernard King and sometimes as Benjamin King. To skirt the issue, writers find it safest to simply call him Captain B. King. The date of the discovery has not appeared in books and it certainly hasn't been celebrated.

The date of the King gusher can be found in Col. Harkness' unpublished manuscript *Makers of Oil History*, 1850 - 1880. When rifling through the Harkness paper recently, Charlie Fairbank unearthed this newspaper account in *The Globe*, dated January 15, 1867: "Captain B. King, Manager of the North Eastern Oil Company brought in a big well on November, 23, 1866, called 'The Big Well' in lot 11, concession XI. It was located by Mr. Kelsey of Buffalo, a celebrated 'oil smeller'."

An abbreviated version describing this date also appears in an article Col. Harkness wrote in the periodical, *Canadian Oil and Gas Industries*, dated February-March, 1951. In his 1940 manuscript, Harkness acknowledges "the references to the 'King wells' are confusing as Captain King drilled a number of wells that made headlines."

The date of November 23, 1866 for the big King well is supported, however, by a written report by J.H. Fairbank in the 1908 Old Boys Reunion souvenir. "Late in the year (1866) the 'King' well was struck..." It's worth remembering that J.H. was living in Petrolia when the King made his terrific discovery. If J.H. kept a diary of this time period it could have shed light on the King discovery, however, no diary has been found covering 1865 and 1866.

Everyone had laughed at Hugh Nixon Shaw, and they laughed again at Captain King, that is, until his well came in. Oil drillers "were convinced oil could be found only near the creeks and rivers," wrote Victor Lauriston in his book, *Lambton County's 100 Years, 1849 - 1949*. King was drilling in the swamp west of what is now known as the aptly named Eureka Street. King had money, Lauriston notes, "But he experienced all the other difficulties that beset an inexperienced man drilling in a new field. Most disconcerting were the unsympathetic comments of the more experienced oil men on the utter idiocy of drilling for oil in such an impossible location."

Today this land is found between Petrolia Line (formerly known as the 10th line) and Discovery Line (formerly known as the Blind Line). The lot number places it west of Eureka Street, on a property currently owned by oil producer Ken Girard.

"Most disconcerting were the unsympathetic comments of the more experienced oil men on the utter idiocy of drilling for oil in such an impossible location."

"The line paid for itself in about eight months of operation and must have been one of the profitable railway branches ever built in Canada."

There is no historical marker to identify the site and few in Petrolia today even know who King was. During the 1950s, most townspeople knew exactly where King's gusher had been. And this is the case with much of the oil history in the area. Facts and stories that were once common knowledge have been forgotten or ignored. When these are learned anew people are amazed and surprised.

The discovery of the King well was obviously good news to anyone in the oil business that November of 1866. And on Christmas day the Fairbanks had another reason to rejoice. The cries of their newborn son, Francis Irving, filled their home. By August there was no baby crying and for Edna, her son's death at nine months old seemed unbearable.

J.H. and Edna went on to have three more children within the next four years. Out of all four births in Petrolia, only one child would survive beyond a year. Francis Irving's death in August 1867 was followed by Huron Hope's death in August 1868, at the age of two months. Mary Edna, born in 1869, would be the survivor and was known as "May" all her life. Their sixth and final child was Ella Leonora, born in 1871 and buried at eight months. It is not known why these three Petrolia babies perished but Edna's letters indicate they had not been robust at birth and their "pinched faces" were a constant source of worry.

In his business life, J.H.'s endeavours flourished. Over the years, he became a respected leader for his quiet integrity and sound judgements, and a pillar in the Petrolia community. In 1866, J.H. was among a band of Petrolia businessmen who desperately wanted rail service in Petrolia. It would be far quicker and easier for transporting oil than dealing with the dreaded mud road to Wyoming.

Those in power at Great Western Railway were not interested in financing a five-mile spur line connecting Petrolia to Wyoming. They might have feared the Petrolia oil boom would be short-lived. J.H. and the businessmen decided to build and finance the spur line themselves. On Jan. 1, 1867 the spur line was fully opened. Immediately, it was wildly successful and Great Western Railway soon purchased it. "This line paid for itself in about eight months of operation and must have been one of the most profitable railway branches ever built in Canada," Phelps stated in his thesis.

The Coming of Canada's First Pipelines

History has illustrated that any new technology is greeted with both shouts of elation and screams of despair. This was certainly true in the progress of the oil industry. Hundreds of men had been earning their livelihood transporting oil by horse and they faced the threat of a new technology - the pipeline.

Like the British whale oil companies that tried to stop gas lighting by saying it was 'dangerous, poisonous – or a defiance of Almighty God', the men with teamsters were horrified by the word "pipeline". In both cases, stopping the progress was as futile as trying to put toothpaste back in the tube.

The pipelines were another answer to the deplorable road conditions oil producers faced when transporting their product. Surprisingly little has been written about the oil pipelines in Petrolia and Oil Springs, considering they were the first in Canada.

It is known that the first Petrolia pipelines were built in the 1870s. Col. Harkness was able to shed more light on the different stages of development in his manuscript, *Makers of Oil History*, 1850 -1880. He cites an article in *The Globe*, November 17, 1875, that says the road from Petrolia to Marthaville, a scant few miles to the north, had become impassable to the point where no oil was being hauled at all. "The Blind Line (the road now known as The Discovery Line) ...is constantly mentioned as impassable. The cost of upkeep was more than the municipality could undertake. From the railway sidings and refineries to Marthaville, it had been drained, graded, gravelled, and partially planked, but in wet weather the traffic was greater than the road was designed to carry, and it failed.

"The situation called for a system of main and collecting lines from the wells to the refineries or sidings," wrote Harkness. "These came, but very gradually, collecting lines came first, bringing oil from large and remote wells to the tanks along the main road. Competition was so keen between producers that they were hesitant about launching a combined venture in a main pipeline."

The 1875 account in *The Globe* reported that two Petrolia men, Vanalstyne and Smith, were laying a pipe from what was then known as 'The Great Geyser some distance west of Marthaville. The distance would have been about two miles. "The dispatch

Hundreds of men with teams of horses made their living hauling oil to the railhead. They were horrified by the word "pipeline".

further says that this pipeline and that of (John) Noble's and McDonald's were the only common carriers in the district." Harkness continued. "The statement is also made that the teamsters in the district were very displeased over the proposed project.

"These great pipes extend over six miles from Petrolia and their total length with ramifications can scarcely be less than 20 to 24 miles."

"Transporting oil by horse-drawn wagon undoubtedly reached its peak in 1881," stated Harkness. "A *Globe* special correspondent (September 3, 1881) says, 'The removal of the oil to the railway station and refineries from a distance which at its further limits extends nine miles from Petrolia, is a work of great magnitude and furnishes a living for several hundred people...The delivery by wagon tanks, great as it is, does not equal the delivery through great pipelines extending to most of the large oil lots...These great pipes extend over six miles from Petrolia, and their total length with ramifications can scarcely be less than 20 to 24 miles."

Pipelines were also making news in Oil Springs. New oil was found in Oil Springs in 1881 and the following year there was enough optimism for the Petrolia Crude and Tanking Company to build a pipeline from Oil Springs to Petrolia. In 1883, a second pipeline to Petrolia was built from Oil Springs, this one laid by the Crown Warehouse Company.

The teamsters' resistance to pipelines was in vain. The network of pipelines continued to grow throughout Petrolia's oil boom. After Imperial Oil set up its refinery in Sarnia in 1897, oil producers would take their crude to Petrolia receiving stations and from there, it was sent by underground pipelines to Sarnia. "These short lines, together with gathering lines in the fields, comprise all of the oil pipelines in Canada in 1912," notes Christopher Andreae in *Lambton's Industrial Heritage.*

Oil Producers Unite

While these changes in the transport of oil were underway, a power struggle was developing between oil producers and oil refiners. In 1877, oil producers in Petrolia attempted to regulate oil prices and for the first time, they organized. On October 24, 1877, they formed a group called The Mutual Oil Association and immediately hiked the price of oil from $1.10 a barrel to $2.08, according the Phelps' thesis.

Refiners did not take kindly to this and they responded by buying very little and using up the oil they had previously stored. When the oil producers had little opportunity to sell their oil they agreed to a lower price. Still, the producers' surplus of oil reached disastrous proportions on Black Friday – May 1, 1879. The price of crude slid to a rock bottom price of 40 cents per barrel. Infighting among the producers quickly led to the dissolution of The Mutual Oil Association.

"When the Association's books were closed," wrote Phelps, "J.H. Fairbank, never at a loss for money, bought 11,000 barrels of crude oil that had been collected from the members as a reserve."

He was not happy. The day after Black Friday, *The Petrolia Advertiser* quotes him saying, "If it was the wish of the producers to have an open market, by all means let us try it. But if after a while they find out their mistake, they must blame no one but themselves for the disaster that may follow."

By 1879, there were eight oil refineries in Petrolia. That year, Jacob (Jake) Englehart added its ninth and most impressive. Englehart was a highly successful refiner and businessman who had arrived in Petrolia in 1870. Partly inspired by tax relief from the town of Petrolia, Englehart moved his refining operation from London and opened The Silver Star, the world's largest and most sophisticated refinery. It was massive and processed 75,000 gallons of crude at a time.

Conflicts between oil refiners and oil producers subsided somewhat and after a few years of floundering, the producers tried uniting again and the Petrolia Oil Exchange was established December 23, 1884. Although prices continued to be low, the Exchange did set a standard for its oil and crude was inspected for its quality. The Exchange, with J.H. as president, started with 34 producers and refiners. More came aboard.

Petrolia boasted nine refineries. Englehart's Silver Star was the world's largest and most sophisticated.

The Imperial Oil Connection

Englehart's move to Petrolia helped centralize the oil industry to a greater degree and smoothed the way for the town to its place as Canada's oil capital. Up until this time, much of the refining of oil had been done in London. On April 30, 1880, a group of 15 London oil refiners banded together with four other refiners to form Imperial Oil. Englehart and Frank Smith were the only refiners from Petrolia; Francis Ward was a refiner from Wyoming and Isaac Guggenheim was a refiner in New York City.

Imperial Oil was formed for several

Imperial Oil's refinery in Petrolia sprawled over 50 acres and could store 100,000 barrels of crude. It also manufactured Canada's only cans.

purposes, the main reason was to retain Canadian ownership and avoid being swallowed by J.D. Rockefeller's company south of the border, Standard Oil. His company had become organized, huge and powerful.

Eight years earlier, smaller American refiners were so fearful of Rockefeller that they took their case to a Congressional Committee according to *The American Petroleum Industry* by Williamson Daum. Quoting from 1872, Daum writes of one American refiner who told Rockefeller he didn't want to sell. "Rockefeller's answer was, 'You can never make money, in my judgement. You can't compete with Standard. We have all the large refineries now. If you refuse to sell, it will end in your being crushed." By banding together, the 19 Canadian refiners had hoped they could stand up to Rockefeller's formidable might.

Imperial Oil would exert enormous influence on Petrolia's development as pointed out succinctly in Petrolia 1866-1966 by Charles Whipp and Edward Phelps: "In 1883, Imperial moved its barrel plant to Petrolia, along with a payroll of $400 a week (at a time when a man made a dollar a day.) ...In December, 1884, the company's head office moved from London to Centre Street, Petrolia...Imperial Oil thus began as a consolidation of outside refineries and made Petrolia by far the major refining centre in Canada, in addition to be almost the only oil producing centre."

There were three key reasons for the London refiners wanting to move to Petrolia. One was that it already had the country's largest refinery, Englehart's Silver Star. Secondly, transporting crude was expensive and London City Council had refused the refiners' request of a $20,000 pipeline from Petrolia. The third reason for the move was that Imperial Oil's London refinery was badly damaged in 1883 and it did not make economic sense to rebuild it.

Imperial Oil's refinery in Petrolia sprawled over 50 acres and could store 100,000 barrels of crude. In its six major stills it produced paraffin waxes, lubricating oils, kerosene and greases. It also manufactured Canada's only cans. They held 10 gallons and were shipped to countries across the Pacific as well as throughout South America.

Despite its size, American petroleum products were still being shipped into Canada. To compete, Imperial enlarged its Petrolia capacity in 1893 with even more products and had 23 branch offices throughout Canada. Two years later it was seeking financial help in Canada and England. None came. Earlier overtures made by Standard Oil had been refused by Imperial Oil but on July 1, 1898 a deal was struck. Standard Oil would open the purse strings for expanding in Canada and Imperial Oil would hand over the majority of its interests. Standard oil affiliates, including the Bushnell Company with refineries in Petrolia, Sarnia and London, were all rolled into Imperial Oil.

The final blow to Petrolia came on February 23, 1889. It's the date "an expanded and strengthened Imperial Oil Company Limited took possession of the Bushnell refinery at Sarnia which was to become - and still is – Canada's greatest refinery," states Gordon A. Purdy in *Petroleum, Prehistoric to Petrochemicals*. Imperial Oil took its operations and its head office out of Petrolia and into Sarnia. Imperial Oil adopted the name "Esso", which comes from the first two letters of Standard Oil. Petrolia's economy and pride staggered from the loss.

Undoubtedly, it was galling for Petrolians to see that Sarnia would call itself The Imperial City. (Sarnia officially became the Imperial City on May 7, 1914. It was the day of the imperial visit by the Duke of Connaught and his daughter, Princess Patricia, when Sarnia officially became a city. Some would say Imperial City commemorates the royal visit. Others would say it was really to honour Imperial Oil.)

The enormity of these developments is best understood when put in context of the times. *Belden's Historical Atlas* captured it well in 1880 by stating: "The alpha and omega of Petrolea is oil, oil, oil. Everything smells of oil; everything tastes of oil; everything is covered and smeared with oil; everything is oil. You hear nothing but oil spoken of in the cars, in the hotels, in the public offices, in the stores, in the Exchange, on the streets, everywhere; and we would think from a casual visit that not only the prosperity of Petrolea, but the lives of all its inhabitants, and the existence of the whole country, depended on whether 'crude' advanced or declined one-eighth of a cent 'on Change' within the next ten days."

Imperial Oil adopted the name "Esso", which comes from the first two letters of Standard Oil.

The Fairbank Family, Supplying Crude to Imperial Oil Since 1880

It has long been speculated that the Fairbank family has been supplying crude oil to Imperial Oil longer than anyone. However, there were so many refineries in Petrolia and Oil Springs that there was certainly room for doubt. Also, J.H. had his own refining company, Home Oil until 1881 and he kept the Fairbank, Rogers and Company refinery until 1897. A found document now dispels that uncertainty and cements this oil family's place in history as Imperial Oil's longest running supplier.

The Fairbank family has been supplying Imperial Oil with crude since 1880, the year gigantic corportation humbly began in London, Ontario.

A "memo of purchase" dated August 12, 1880, bearing the letterhead "Imperial Oil Co., London, Ont." has been unearthed in the family papers. It is addressed to J.H. Fairbank Esquire. It states in part: "Two tanks crude oil. 4,700 barrels. Price $1.58 per barrel delivered at Imp. Oil big tank at Petrolea. Quantity guaranteed. Terms 30 days."

In smaller letterhead at the top it says "J.L. Englehart & Co. Mutual Refining Co." as well as "London Oil Refining Co., Waterman Bros.". Englehart, who has been dubbed "The Father of Imperial Oil", would become its second vice-president.

What's interesting about this memo of purchase is the date. Imperial Oil was first formed on April 30, 1880. This memo of purchase addressed to J.H. is dated August 12, less than four months later. It is also before Imperial Oil was incorporated. That happened Sept. 8, 1880. At the time, J.H. was the largest oil producer in Canada and Englehart was its largest refiner.

The business relation J.H. began with Imperial Oil in 1880 has continued unbroken for more than 120 years. Fairbank Oil continues to ship its crude to Imperial Oil for refining.

Buying Into Oil Springs for the Second Time

While Petrolia boomed after the discovery of the King well, Oil Springs dwindled. The population plummeted from 4,000 to a mere 250 souls. A man named Henry Shannon still held faith in his oil fields and had held on to 138 acres in Enniskillen Township even though they had become unproductive.

Then, in 1881, oil prices rose out of a 15-year slump. It finally meant that oil could produce decent profits again and it renewed interest in the wells of Oil Springs. The president of the Excelsior Oil Company, W.S. Duggan, found there was lots of oil left in Oil Springs, you just had to drill deeper for it. This oil was accessed at 400 feet, a full 150 feet deeper than where the gushers were found.

Oil Springs surprised everyone with this second boom. It was definitely quieter than the 1860s, but the oil was flowing once again. There was enough optimism in 1882 to extend the Canada Southern Railway from Courtright on the St. Clair River eastward to Oil City, only two miles north of Oil Springs.

Also in 1882, the Petrolia Crude Oil and Tanking Company built a pipeline connecting the Oil Springs field to its tanks in Petrolia, according to Harkness. "...and the Crown Warehouse Company duplicated this pipeline in the following year."

With good oil prices it was not surprisingly that when Shannon was ready to sell his land, J.H. was ready to listen. "I have no doubt it is the best oil property in Canada," Shannon wrote to J.H. in September, 1882. "I believe the quantity of oil unlimited in that land." By November, the deal was sealed; J.H. bought a two-thirds interest from Shannon for $16,000, agreeing to develop the property and regularly remit a share of the proceeds to him.

Shannon had two partners, W.S. Brown and B.B. Brown, and they too finally agreed to the sale of the remaining one-third. "I have no desire to dispose of my interest," W.S. Brown had written to Shannon a month earlier, "as I would much prefer to retain it under the very able and honest management of Mr. Fairbank."

In parting with the remaining one-third of the property Shannon sounded wistful. He had purchased the land before the first Oil Springs boom and it represented much more than just a source of income to him.

"It is 30 years this fall that I first became interested in Lot 18 in the 2nd concession of Enniskillen, and, as you well know, that much of that long time that property has been part of myself; the hopes, anxieties, fears and success of that 30 years in my early life in connection with this subject seems to me to have been part of a beautiful dream," he wrote in a letter to J.H. in September 1890. "I can say with gratitude, that the property has been a great pleasure to the writer. Yet all material things should be treated in a material way..."

It's clear that J.H. was intent on developing the Shannon land to its full potential. The Fairbank papers show he infused the property with $32,000 between 1882 and 1887, a sizeable sum in the 19th century. It was good news for J.H. that rail arrived in Oil Springs in 1886; a spur line was built to link Oil Springs to Oil City.

Sometime in the 1890s, J.H. acquired

"...that property has been part of myself; the hopes, anxieties, fears and success of that 30 years...seems to me, to have been part of a beautiful dream."

"Nine-tenths of the oil in this country is manufactured here now."

an adjacent parcel of land to the south and to this day this section of Fairbank Oil is known as 'The James Property'. It had, and still has, a house on it that was likely built during the late 1880s. Records show J.H. also held seven other sites with wells on them at this time but now that a century has passed these names have lost their meaning and can no longer be tied to exact locations. It is known that Fairbank Oil comprised some 160 acres at the turn of the century.

An amazing amount of family records have been kept, yet, there are pieces of the puzzle that may never be found. It is not clear which properties J.H. owned in Oil Springs between the late 1860s and the early 1880s. In his lifetime, J.H bought and sold more than 300 pieces of real estate, according to Phelps' thesis.

The Ending of the 19th Century

The second boom of Oil Springs was important, however, it had been totalled eclipsed by Petrolia's spectacular developments. J.H. was very clear in his statements to the 1890 Royal Commission on how important Petrolia had become. "Nine-tenths of the oil in this country is manufactured here now, and that has increased the number of men very considerably. There are large quantities of chemicals used here, our boilers are manufactured here, our stills are made here, and our brass goods are partly made here. All our engines are manufactured in Canada...We make as good a lubricator as ever was manufactured. I have used it for five or six years, running engines, and they have had no repairs. I prefer it to the best lard oil. I supply it to the Lake Superior steamers, and they find they can use it in their steam cylinders."

Another Petrolia oil man, Martin Woodward, enlightened the Royal Commission on illuminating oil. Three grades were manufactured from Lambton crude – Homelight, Economy and Standard. (These are the same names given to different grades of light bulbs today.) "Our oil is better in some qualities than the American," he told the commission. "It gives very much the same flame, and of the same colour, but is not so free from sulphur compounds as the best American."

Through the 1880s all of Petrolia was bustling. The town's economy was branching out and a new era of manufacturing was emerging – the Stevenson Boiler Works, the creamery, the Petrolia

Wagon Works, the precursors to Oil Well Supply, the Pork Packing Plant. It was a time of much building – the downtown blocks, homes, Victoria Hall, and churches.

While these developments were occurring in oil business, J.H. and Edna's three children - Henry Addington, Charles Oliver and May - were growing up. Though J.H. was largely self-educated, he made certain his sons would receive excellent formal education. In the 1880s, when a university education was very rare, the eldest son, Henry, had graduated from the University of Toronto and was studying either medicine or pharmacy at the University of Michigan in Ann Arbor.

Charles studied at Helmuth College in London, which later became part of the University of Western Ontario. To his father's delight, Charles then studied at the newly established Royal Military College in Kingston, graduating four years later in 1880 as a lieutenant. (This is well known to anyone who has studied at RMC since that time. The college has a long-standing tradition that all graduates must memorize and recite the 18 names of the original class of 1876.) After graduating, Charles obtained a commission in the Royal Artillery (England) and went to Woolwich Academy in England for further training in the riding school and arsenal before joining his battery.

The eldest son, Henry Addington, died suddenly at the age of 24. It was a profound shock to the family. He committed suicide in February, 1881 following an entanglement with a Canadian woman. Charles quickly returned from England to be closer to his grieving parents and sister, May. He was now J.H. and Edna's only son. Shortly after returning to Canada, Charles took a commission in London, Ontario Field Battery.

Charles was now J.H. and Edna's only surviving son.

In 1888, he began a six-year stay in New York City. There, he entered the College and Physicians and Surgeons and then studied at Columbia University and received his medical degree in 1891. Charles continued post-graduate work until 1894. He returned to Petrolia afterward, but, for reasons unclear, he never made medicine his full-time profession or tried to obtain a licence to practise on this side of the border. His later diaries show that his medical advice was often sought and he did assist in some surgeries. One entry mentions an oil accident in Bothwell where he helped amputate a man's leg.

When in Petrolia, he worked closely with his father in their many businesses. It appears that they were always close. It was Charles, not his older brother Henry, who had come to live with J.H. in Oil Springs as a four-year old. His mother, Edna, and brother, Henry, arrived in Petrolia three years later. Though Charles would later oversee the work in Oil Springs, oil was not the singular obsession for him that it had been for his father. He was interested, not only in medicine, but also pursued politics and a military career. In fact, he was known by three names; Major Fairbank, Dr. Fairbank, and "The Little Doctor" because he was short in stature.

Charles was interested, not only in medicine, but also pursued politics and a military career.

The closing 15 years of the 19th century would see the wealth of Petrolia set out in architectural splendour. Smart brick buildings lined the main street, replacing modest wooden structures. The first of these was the Archer Block and it was here that the Petrolea Club was established for reading the latest periodicals and playing billiards. Beautiful churches were erected and many gracious homes were added.

The grand Victoria Hall opened in the heart of the town in 1889. Designed by George Durand, one of the top Ontario architects of the day, it boasted a stately clock tower, exquisite stained glass windows and ornate wrought iron railings along the balcony. It housed the town council chambers, the fire department, the town jail and a large hall with a stage that proved excellent for touring performers of all sorts. Four times each year, the elite would dress in their finery for the formal balls, called the Petrolia Assemblies, at Victoria Hall.

The Fairbank mansion on the town's main street was elegant, understated and immense, the largest in the county according to Phelp's thesis. It was completed and ready for the family by 1891. Petrolia Assemblies were occasionally held in the Fairbank mansion's third floor ballroom.

The mansion was officially called Sunnyside and the eight bedrooms were far more than the family required for its own use. J.H., Edna and their daughter, May, were the only ones to live in it year-round. Henry had died in 1881 and the second son, Charles, spent most of his time between 1888 and 1894 studying in New York City. (See Asides, pg. 147)

Earlier, Charles had joined the 6th London Field Battery and each year he trained for a few weeks in London

(Ontario). In 1888, before leaving to study in New York, he took command of the battery. When Britain declared the Boer War in 1899, Charles spent a good deal of his time recruiting for the South African conflict. English Canada rallied behind the cause, and more than 6,000 men enlisted before victory was declared on May 31, 1902. Returning men were welcomed home with great parades, the 244 casualties were honoured with monuments. For Charles and so many others, the Boer War set the stage for their later involvement in World War I.

Despite the wealth, comfort and cultural life that surrounded Edna, the 1890s were not happy years for her. After living only three years in the mansion, she retired to Pasadena, California with her daughter, May, who was then 25 years old.

The first suggestion of California can be found in an 1877 letter to Edna. Her cousin Sarah White wrote, "Mother mentioned your possible trial of treatment in Chicago. Let me prescribe Southern California – a climate that dissipates every ill that flesh is heir to." This proved to be important, for a whole branch of the family would start their roots in California after Edna's need to go there for her health. Various members of the Petrolia family would buy land and build homes there. They even purchased the Elk Hills oil field in California, a property that would have made them fantastically rich if they had been able to hang on to it.

Edna outlived four of her six children and survived long enough to see her daughter, May, married to Huron Rock in California on New Year's Day, 1896. Two months later, Edna died in Pasadena at the age of 66, never seeing her first grandchild, John Fairbank Rock, born that December.

"Let me prescribe Southern California - a climate that dissipates every ill that flesh is heir to."

John Henry Fairbank – The Entrepreneur

In addition to his oil properties and owning the largest hardware store west of Toronto, J.H. created and owned the town's first real bank with his partner, Leonard Baldwin Vaughn in 1869. This is the same Vaughn who had produced a vast amount of oil and excitement nine years earlier in Oil Springs. Officially it was called Vaughn & Fairbank, Bankers, but everyone knew it as The Little Red Bank. The two men had operated a bank together in Oil Springs and moved the business and the building to Petrolia. It opened August 10,

Altogether, J.H. Fairbank employed more than 400 people in his business empire.

1869 and when it closed 55 years later in 1924, it had become one of the last private banks in Canada. Presently, it is the office of lawyer, Wallace B. Lang and it has the distinction of being Petrolia's oldest surviving building.

By the 1880s Petrolia was positively bustling but J.H. was once again looking ahead. He was not sure how long oil would dominate the Lambton County economy and thought its long-term future may be in agriculture. In fact, he had been buying farmland for years.

As a young man, he had farmed for several years in Niagara Falls and knew well that hard work didn't necessarily translate into profit. Drawing on the banking experience he gained from opening the Little Red Bank in 1869, he then led the opening of another financial institution, The Crown Savings and Loan. It was chiefly devoted to granting mortgages to good farmland and had several successful farmers among its directors. Incorporated in January 1882, J.H. would serve as its president for 30 years. Englehart was vice-president for 25 years. (The building of Crown Savings and Loan on Petrolia's main street later became Industrial Mortgage and Trust, then Royal Trust and is now the Royal Bank.)

In addition to his two banks, hardware store and Fairbank Oil in Oil Springs, J.H. also owned two refineries; Home Oil Company and Fairbank, Rogers & Company; and had interests in railways and lumber. In 1891 he bought the Stevenson Boiler Works at a sheriff's sale as well. Altogether, he employed more than 400 people in his business empire. J.H. once held one-quarter of Petrolia's real estate according to Phelps.

The Stevenson Boiler Works was a significant industry in Petrolia. It employed 55 people and was the largest business not directly related to oil. The owner, William Stevenson, had slinked out of town owing Crown Savings and Loan $14,000 and had a number of other unpaid loans. J.H. bought it at the sheriff's sale for $14,000 and became keenly interested in its operation.

His daughter May took it upon herself to inform her brother, Charles, of the new development while he was away at medical school. She wrote, "Stevenson, boiler maker, has sought a more congenial clime; He forgot to pay some $60,000 he owed before leaving, but that is a slight matter when he needed an immediate change of air...Papa has a mortgage on the boiler shop and has taken possession; intends going right on with the work. He talks of nothing but boilers; told Mama she needed a new flue-sheet and a patch on left side of firebox, then he went about hammering stoves, etc. to see if they were sound. We have been having lots of fun about it, might just as well, you know, but Papa looks tired and worried, says he wishes I was a boy – now I don't."

Petrolia's newspaper, *The Advertiser*, which was openly opposed to J.H. politically, heralded him as something of a hero: "Thus through the foresight and characteristic enterprise of Mr. Fairbank, what would have undoubtedly proved a severe loss to Petrolia, will, through his efforts be made a bigger and more permanent benefit than ever."

A report in *The Globe* on June 24, 1893 made it clear what Stevenson Boiler Plant meant to Petrolia. "The amount of business done will be more easily understood when it is stated that more than 90 per cent of all the boilers, tanks, stills and good requiring heavy plate that are used in the production and refining of oil are made at these works. Besides the local trade, they do a large trade in marine, stationary, and portable boilers, which are finding their way into all parts of the world."

The boiler works was refitted and continued as a profitable business. Just before the turn of the century, it began making fire extinguishers too. They didn't work well, however, and in 1904, J.H. modified it and it became known as The Fairbank Fire Extinguisher and sold well.

He must have spent an inordinate amount of time at meetings and was a strong voice in oil producer associations fighting to get oil prices stabilized. Refiners were gaining power over the oil producers and to combat this, J.H. and other successful producers banded together and opened their own refinery, Home Oil Company, in 1873. For the next eight years, J.H. was its president and general manager. The refinery occupied 16 acres and could refine 3,000 barrels of crude each week. It was sold to Imperial Oil in January 1881.

Over the years he held a multitude of board positions apart from the oil business. Among them were: head of the cemetery committee, the building committee for Victoria Hall, and head of the clean water committee. He was also reeve for three terms and a federal Liberal member of parliament from 1882 to 1887.

The Stevenson Boiler Works made more than 90 per cent of all the boilers, tanks and stills used in producing and refining oil.

"...everything in and of the town is the most inflammable material."

J.H. was even chief of the Fire Brigade for three terms (1874 to 1889) and in Petrolia that meant something. The Belden Atlas of 1880 noted the town's fire department was "first class" and added, "and in no town is this attribute more required...considering that everything in and of the town is of the most inflammable material."

John Henry Fairbank

This rare photograph shows J.H. as a young man. He was 29 years old when he first arrived in Oil Springs in 1861. His son Henry was four years old and Charles was approaching his second birthday. At that time, they stayed with their mother, Edna, on the farm in Niagara Falls.

Fairbank Family photo

First Home in Petrolia

In 1865, J.H. built his first home in Petrolia, on the site where he would later build his mansion. It was one of four homes west of Bear Creek at that time. Before construction of the mansion, it was moved to the north of the lot and served as a carriage house. It remains there today.

Fairbank Family photo

Second Home in Petrolia

The second Fairbank home in Petrolia was slightly grander. It stood to the west of the first home. At the left of the photo is the home of oil producer John D. Noble. It stood at the northeast intersection of Tank Street and Petrolia Line.

Fairbank Family photo

The Fairbank Mansion

The Fairbank mansion took two years to build and was completed in 1891 on the main street of Petrolia. It astounded all with its enormity and the detailed workmanship inside it. J.H. Fairbank lived in the home for 23 years before it passed to his son Charles.

Fairbank Family photo

The Family Gathers

This photograph was taken on January 1, 1896 in Pasadena, California. The occasion was the wedding of J.H.'s daughter, May, to Huron Rock. J.H. stands second from the right, and his wife Edna, is seated in the dark dress next to him. At the far left is his son, Charles. Others in the photo are unknown.

Fairbank Family photo

A Day at the Beach

This photograph was taken about 1910, possibly by J.H's son Charles. Standing at right is J.H., and standing at left is his grandson, Charles II (later known as Charles Sr.). Seated from left to right are: his grandsons John Henry Junior, and Henry Churchill; and their mother, Clara. Others in the photo are unknown.

Fairbank Family photo

John Henry Fairbank and Grandson, Charles II

Posing outside the Fairbank mansion are J.H. and his grandson Charles II. This photo was likely taken around 1910. At the edge, to the right, two other grandsons, John Henry Junior and Henry Churchill are partly visible.

Fairbank Family photo

Employees of John Henry Fairbank

As a Christmas gift in 1895, the employees of John Henry Fairbank presented him with this group photograph. For more than 100 years the photograph has been hung in VanTuyl and Fairbank Hardware Store. It remains there today.

Fairbank Family photo

John Henry in his Mansion

The Fairbank mansion has been called the last great creative act of J.H.'s long career. He is pictured here beside the grand staircase made of rubbed sycamore cut from Fairbank farms in Brooke Township and aged for two years before construction. The date of the photo is unknown but likely around 1910.

Fairbank Family photo

London Oil Refining Co
Waterman Bros.

REFINED OILS
HIGH TEST
PARAFINE CANDLES & WAX
AMBER OILS
BOLT & SCREW CUTTING OIL
WOOL SPINDLE
SEWING MACHINE & ENGINE OILS
OLEINE
RAILWAY & BLACK OILS

J. L. Englehart & Co
Mutual Oil Refining Co

The Imperial Oil Co.
London, Ont.
Aug 12th 1880

Memo of Purchase from
J. H. Fairbank Esq.

Two Tanks Crude Oil 4700 Barrels price $1.58 per Barrel delivered at Imp Oil Co's Tanks at Petrolea, quantity guaranteed
Terms 30 days.
When pumped out, the Pump to be placed at 31st line at Bottom.
Tank rent 5c per Barrel per annum, to be paid from this date by I.O.C. until Oil is deld.

Selling Crude to Imperial Oil in 1880

The Fairbank family has been supplying Imperial Oil with crude longer than any other producer. The famous company was formed in London, Ontario on April 30, 1880. The one "memo of purchase" to J.H. is dated August 12, 1880. The second document, dated December 31, 1880, shows how the letterhead changed after the company was incorporated in September that year.

Fairbank Family papers

London Oil Refining Co.
Waterman Bros.

J. L. Englehart & Co.
Mutual Oil Refining Co.

Office of The Imperial Oil Company
(LIMITED)
Producers, Refiners & Shippers of Petroleum

WORKS. London & Petrolea.
HEAD OFFICE. London.

London, Ont. Dec 31st 1880

J. H. Fairbank Esq.
Dear Sir
Kindly deliver to us the oil in the "Noble" tanks & oblige

THE IMPERIAL OIL CO[illegible]
(LIMITED)
per J. M. Fowler

Chapter 3 – The Second Generation: The Era of Dr. Charles Fairbank

After the death of Edna in 1896, J.H. and his only remaining son, Charles, continued to live and prosper in Petrolia. The two proved to be excellent business partners together. They even shared the same birthday of July 21. Charles' military career would be an intermittent one and the business of oil would be more continuous. In addition to the Oil Springs field, the Fairbank family also discovered and developed new oil in Bothwell, 20 miles to the southeast.

The Bothwell oil fields had already boomed once and this discovery created the area's second boom. Charles bought land there in 1898 with his partner Frank Carmen. Today, anyone looking closely at the 1907 steam engine in the rig (the powerhouse) will see the initials of the two partners. The diaries of Charles show that he was frequently hopping the train from Petrolia to Bothwell to attend to oil business there. And it was in Bothwell where he met the woman who would become his bride.

The year the new century was ushered in Charles turned 42 and married Clara Sussex of Bothwell, then 23. (Clara also used the names Clarissa and Clare). Since J.H. was living in the mansion, the couple bought a large, handsome home in Petrolia's Crescent Park neighbourhood that was known as The VanTuyl House or The Yellow House (420 Warren Avenue). This home had belonged to Benjamin VanTuyl, who died in 1900. VanTuyl had been a friend and valued partner of J.H. who managed his hardware store for 26 years. As a measure of respect and gratitude, VanTuyl's name remains with the store and to this day it is called VanTuyl and Fairbank Hardware.

Charles and Clara went on to have four sons: John Henry II in 1901; Charles Oliver II in 1904; Henry Churchill in 1906; and nine years later, their fourth

Charles was educated as a doctor but his colorful career would be in oil, politics and the military.

son, Robert Theodore, was born. Although they were based in Petrolia, the family would be spending a good deal of time in California, in fact, their first son, John Henry II, was born in Santa Barbara. This explains why the Fairbank name was attached to an oil field in California as well as a gas gusher in Oil Springs.

The Fairbank name was attached to an oil field in California as well as a gas gusher in Oil Springs.

In 1913, a syndicate of producers pooled their resources and decided to drill deeper in Oil Springs. They reasoned that deeper drilling created a second boom in the 1880s and perhaps a third boom could be created. The first deep test was on an acre of ground donated by Charles Fairbank. Difficulties set in and it was abandoned. A second test was abandoned too. While a third test was underway, Charles independently tried again. The Fairbank gusher came in March 7, 1914 yielding not oil, but gas. At first, it produced 2 million cubic feet a day and two days later, it hit an impressive 11 million cubic feet daily with 830 pounds of pressure. It had been Lambton County's biggest gas find.

Finding the gas took two months of drilling. It was sold to Union Natural Gas Company of Canada Ltd., and piped to Petrolia. Today, the site can be located on Fairbank Oil Properties, slightly northeast of the blacksmith shop.

The Fairbank gusher's fame would be short. It was played out in two months. By May that year, an even bigger gasser was found in Oil Springs. Just 500 feet away, the Oil Springs Oil and Gas Company, drilled to a depth of 1,906 feet and it was rewarded with 22 million cubic feet of gas daily.

Charles and his partner, Frank Carmen, also bought land in what would become a lucrative section of the Elk Hills oil field. They had bought the land from the State of California in 1908, and then they leased it to Standard Oil Co. of California. The company struck oil in 1919. Following the Teapot Dome political scandal in the 1920s, the U.S. government reclaimed this oil field and those belonging to several others. (see Asides, pg. 181.)

Charles and his wife, Clara, had built a secondary home in Morro Bay, an 11-acre site midway between San Francisco and Los Angeles. Clara had found the site with a scenic view of the Pacific Ocean only because she was riding in a car that had gotten lost. According to Clara, she designed the house. Standing on the undeveloped land, the builder, Tom Bicknell of Petrolia, asked her, "Madam, where would you like your front door?" She then marched to a spot on the grass and pointed. The Morro Bay home would become a centre for the extended family for decades. J.H. visited there often but always kept tabs on his businesses at the same time and was frequently firing off letters from California to his manager, Alex McQueen.

The eldest son of Charles and Clara, John Fairbank Junior, had been born in Santa Barbara before the Morro Bay home had been built. As an adult, he bought a ranch near Morro Bay, at San Luis Obispo, where he raised livestock. His daughter, Claire, would be born there in 1922.

At the age of 81, J.H.'s health started to decline and he had lost a good deal of his hearing two decades before this. In 1912, he handed all his business dealings over to his son, Charles. To assist his father, Charles moved his family into the mansion, selling their home, The VanTuyl House. J.H. lived two more years, and at the age of 83, he died in the mansion on February 10, 1914. It was three weeks before the Fairbank Gusher and less than six months before World War I erupted.

The day of his death, the town businesses shut down and flags were lowered to half-mast. During his life in Lambton County he rose from a penniless surveyor living in a log shanty to the head of a business empire living in a grand mansion. It was a life of success through hard work and his deeds created waves of change, not only for the petroleum industry but the individual lives of many.

Perhaps one of the most telling tributes was, not after his death, but when friends gathered for his 80th birthday. "Your intercourse with your fellow men in business matters has always been characterized by a spirit of brotherly kindness which has made the title of 'J.H.' a mark of high appreciative distinction," said a friend quoted in the Phelps thesis. "You have ever responded to the call of public duty..." The funeral procession for J.H. Fairbank was the longest the town of Petrolia had ever witnessed.

In the summer that followed, Charles resumed his military career with World War One breaking out on August 4, 1914. Initially, he was to recruit for the 70th Battalion but he felt that to do the job effectively he needed to have first-hand knowledge of the conditions. Although he had reached the advanced

The funeral procession for J.H. Fairbank was the longest Petrolia had ever witnessed.

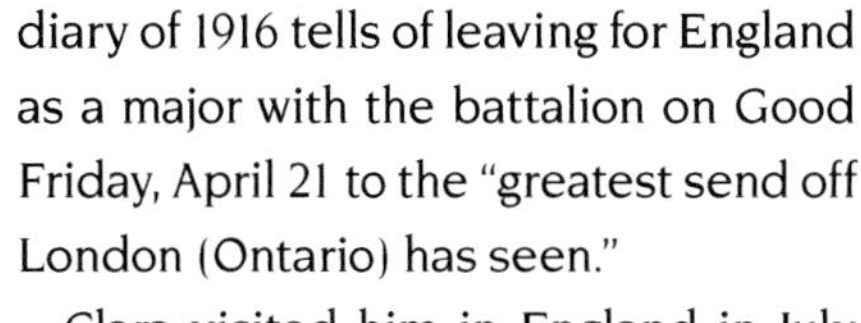

age of 58, he fought in the trenches. His diary of 1916 tells of leaving for England as a major with the battalion on Good Friday, April 21 to the "greatest send off London (Ontario) has seen."

Clara visited him in England in July when the Battle of the Somme was already underway. He had transferred to the 18th Battalion and his diary records that when he got word he was being sent to the front, Clara took the news badly; "(I) never saw her in such a state. I was much alarmed." Arriving in France on August 3, Charles filled many pages of his gilt-edged diary with grisly accounts of the invasion of "the Huns" and bloodbath he witnessed from the trenches at The Battle of the Somme. At the battle's end almost 1.3 million were dead; 24,713 were from Canada and Newfoundland.

Although he was 58 in World War I, Charles fought in the trenches in The Battle of the Somme.

Charles returned home to Clara and his four sons on November 16, 1916. His last entry in his diary of 1916 reflects a horrific year with uncertainty of what suffering may lie ahead. "The old year is gone with its scenes of carnage and desolation and misery. May the new year bring victory to the Allies that will assure peace for the future."

In Canada, he would see more military duty in 1918 when he took a staff position at the Headquarters Military District 1, London, and remained until the end of the war.

Like his father, Charles was interested in politics. Like his military career, his political career would be intermittent and overlapping his many other interests. In 1911, he ran as the Liberal candidate in the federal election but did not win. Later in 1914, when so much else was going on in his life, he was also the Warden for Lambton County. From 1915 to 1918, he was elected reeve and he followed this up by becoming the mayor of Petrolia in 1919.

Charles did not, however, share his father's gift for longevity. Only six years after serving as mayor, he died in Santa Barbara, California at the age of 66. His four sons were still young; John was 23, Charles Oliver II was 21, Henry was 19, and Robert was only nine years old. Clara was made a widow at the relatively young age of 46.

As the family did for his mother, Charles' casket was sent from California to Petrolia by rail. He was given a full military funeral; a firing squad leading the procession, his casket draped in the Union Jack and carried by gun carriage. In a very long obituary on February 25, 1925 the *Petrolia-Advertiser*

Topic said: "His wide experience and well-ordered brain made him a splendid adviser and his opinions were often sought in small things as well as large. He was ever ready to lend assistance. Always honourable and righteous in his living, he became an example for many younger men who admired him for his character, his kindliness and his quiet unostentatious benevolences."

Declining Fortunes

During the 1920s and 1930s the family's wealth diminished considerably. There were several factors, some of them rooted in events at the turn of the century. The story of the Fairbank fortune was interwoven with the epic events of the area and the world. Petrolia's glorious oil boom was over and would never return. Oil fields were no longer gushing; they had slowed to a trickle.

Without oil to lubricate the local economy, the town pinned its hopes on diversifying. One promising business was the Petrolia Wagon Co. Ltd. Starting in 1901, it produced sleighs and various horse-drawn wagons including wagon-wheeled ambulances. It grew quickly and for almost two decades it was the largest industry in Petrolia that was not related to oil.

The Wagon Works may have been large but it was forever mired in a financial mess. The owners turned to J.H. for help. In a move that would cause the greatest loss in his career, he signed a piece of paper on December 29, 1908. It was a document stating that he would guarantee any and all loans the Bank of Toronto chose to give the Wagon Works. J.H. would not live long enough to see the debacle that followed 12 years later.

The owners of the Petrolia Wagon Works turned to J.H. for help. In a move that would cause the greatest loss in his career, J.H. signed a piece of paper on December 29, 1908.

Financially wobbly or not, the wagon works could have never withstood the deathblow to come. It was not foreseen that something called a motor vehicle would soon revolutionize transportation and swiftly render wagons obsolete.

The Petrolia Wagon Works Co. Ltd. collapsed. On June 23, 1920 a letter addressed to Major Charles Fairbank was sent from The Bank of Toronto's head office. "...it is necessary for us to ask for the payment of the balance owing to this bank, amounting to $210,900 with accrued interest." The loss was quite devastating and with one stroke of the pen it wiped out a sizeable portion of the family's wealth. It is unclear which properties the Fairbanks needed to sell to cover this incredible defaulted loan.

"I am glad to note that the flowers, foliage, mountain valleys and ocean are not going to alienate your affections from Enniskillen mud and good-looking people."

The letters of J.H. show that he valiantly tried to detach himself from the Petrolia Wagon Works years before his death in 1914. While visiting his daughter in California in 1910, J.H. was sending a flurry of letters to his business manager Alex McQueen and he was clearly concerned about Mr. English, who headed the Wagon Works. (Note: McQuien would later change the spelling of his name to "McQueen".) In a letter dated May 2, J.H. wrote: "My Dear Mr. McQuien, Do you fully understand and does Mr. English fully understand how important - how absolutely necessary I consider the sale of the Petrolia Wagon Works?...I will not continue to remain in the conundrum I now occupy with it...I speak of 'sale'. What does this mean? It does not mean a sheriff sale in bankruptcy."

The tone of this particular letter is not indicative of the excellent relationship he had with McQueen and how much J.H. respected his judgement. Throughout J.H.'s stay in California, they exchanged many letters and covered a great range of business quite efficiently. There were also warm, chatty bits from McQueen: "I expect to hear daily that you have been out automobiling," he wrote in one April letter. Later that month, he finished a letter to J.H. by writing: "I am glad to note that the flowers, foliage, mountain valleys and ocean are not going to alienate your affections from Enniskillen mud and good-looking people."

Shortly after the death of J.H. in 1914, the family lost the services of McQueen, who so skilfully managed much of the Fairbank business empire. He was wooed by Imperial Oil and became its vice-president in 1916.

More bad news beset the family in 1920s, when the U.S. government took over all lands in the Elk Hills oil field in California. The family had been receiving royalties from the oil found there by Standard Oil Co. of California. Decades later, Robert, the youngest son of Charles and Clara, would spend considerable resources legally contesting the U.S. government, but never succeeding. (See Asides, pg. 181.)

By 1925, Charles had died. Two years later, his eldest son John Henry II, died. He accidentally shot himself while he hosted a Halloween party in his gracious home at 4058 Petrolia Line. (Much later this house would be known as Rebecca's Bed & Breakfast.) He was 26 years old. His only child, Claire, was but 5 years old. Earlier, he had a ranch in California and his daughter was born at San Luis Obispo. The family had moved back to Petrolia because his wife, Eileen, missed her friends. Once they were in Petrolia again, John bought land to the west of town and farmed it.

Four months before the death of John Henry II, there had been a happy event - a small intimate wedding at the Fairbank mansion. On July 2, 1927, Clara, Charles' widow, married Leo Ranney. *The Advertiser Topic* reported "Mr. Ranney is one of the foremost mining engineers of America and is associated with the Standard Oil Co. of New Jersey." He was also well-regarded as a geologist and inventor who came to have 3,000 patents to his name. It is not known how and when the two met. It's assumed that his geology work had connected him with the oil industry. The tumultuous decade of change closed with the stock market crash of 1929 and all prices plummeted during the Great Depression of the 1930s.

Clara was low on funds and in 1932 sold the main street buildings that comprised the major portion of Van-Tuyl and Fairbank Hardware. She also sold the oil properties in Bothwell. Despite these depleting or vanishing sources of income, Clara and her family continued to live in the style they had become accustomed to. The job of keeping Clara Fairbank Ranney in cash would fall to the shoulders of Charles Oliver II, the second of her four sons.

A happy event – a small intimate wedding at the Fairbank mansion.

Henry Addington Fairbank

Henry Addington Fairbank was the first child born to J.H. and Edna. His untimely death at the age of 24 meant that his younger brother Charles would survive as the only son.
Fairbank Family photo

Dr. Charles Oliver Fairbank

Known as Dr. Fairbank, The Little Doctor or Major Fairbank, his various names reflected his wide ranging career. Throughout his life he was a doctor, an oil man, an entrepreneur and a military man. He was also a county warden, mayor of Petrolia and ran as a member of parliament.
Fairbank Family photo

Charles and Clara Fairbank

Charles was in the first graduating class of Canada's Royal Military College and despite his age of 58, he fought in The Battle of the Somme during World War One. Before being sent to the front, Clara visited him in England.

Photo by G.B. Robson, Fairbank Family photo

World War One

After World War One broke out in 1914, Charles initially recruited for the 70th Battalion. This photo was taken from the west side of Fairbank mansion and shows a three-pole derrick to the north.

Fairbank Family photo

Dr. Charles Fairbank

Charles and his wife, Clara, moved into the mansion in 1912 with their three sons to help J.H. in his declining health. Their fourth son was born in the mansion. They also spent considerable time living in California and their eldest son, John Henry Junior, was born in Santa Barbara.

Fairbank Family photo

Clara and Fairbank Sons

Taken about 1911, this photo shows three of the four sons born to Charles and Clara. From left to right, Charles II (later to be known as Charles Sr.), John Henry Junior, Henry Churchill and Clara. The fourth son, Robert, was not born until 1915. In this photograph, John Henry Junior bears an uncanny resemblance to "Little Charlie" Fairbank" in 2003. ("Little Charlie" is the son of Charlie Fairbank and Patricia McGee.)

Fairbank Family photo

Charles and Friends in the Mansion

In this photo, Charles is seated second from the left. Alex McQueen, who was J.H. business manager and later an Imperial Oil executive, is seated at the far right. Clara is seated at the right in the middle row. A photo of J.H. can be seen on the wall. This was probably taken about 1915.

Fairbank Family photo

J.H.'s Letter To Alex McQueen

By guaranteeing an enormous loan to the Petrolia Wagon Works, J.H. feared a huge financial loss. In this 1910 letter to his business manager, Alex McQueen, J.H. urges that the business needs to be sold.

Fairbank Family papers

The Bank of Toronto
Head Office,
Toronto CANADA 23rd June, 1920.

Major C. O. Fairbank,
Petrolia, Ont.

Dear Sir:-

Petrolia Wagon Co. Ltd.

The operations of this Company having practically come to an end, it is necessary for us to ask for payment of the balance owing to this Bank, amounting to about $210,900. with accrued interest.

As you are one of the guarantors to the Bank, we must ask you to take measures to provide for the liquidation of this debt within the next thirty days.

The outcome of this venture has been very unfortunate, and we greatly regret the losses sustained by those interested, but the situation has to be faced and we hope you will give it your prompt attention.

Yours truly,

General Manager.

The Bank Came to Collect

In June of 1920 the Bank of Toronto came to collect after the Petrolia Wagon Works collapsed. Charles Fairbank had 30 days to pay more than $210,000. It wiped out a sizeable portion of the Fairbank fortune.

Fairbank Family papers

A Gathering at the Fairbanks

Before moving to the mansion in 1912, Charles, Clara and family lived in the yellow VanTuyl House on Warren Ave. in Petrolia where this photograph is taken. Clara Fairbank stands near the centre with a large bouquet on her dress. To her right is Charles, and on her left is oil man "Pa" Jenkins. The date and occasion of this shot are unknown.

Fairbank Family photo

Ottawa, 2nd Oct., 1911.

Dear Dr. Fairbank,-

Accept my sincere thanks for your kind letter. As you say, the cause, for which we fought and have been defeated, is sure to revive, because the policy was sound and strong. The people have been bamboozled by appeals to passion and prejudice but they will recover from this panic. We have only to bide our time. I thought it well, under the existing circumstances, to remain at the head of the party, and evidences are accumulating in my hands, from all parts of the country, that, though defeated, the party is still strong.

Let me tell you how much I appreciate the great fight you put up in East Lambton. Let me add that I will be glad at all times if you will favour me with your views on the different public questions as they develop from day to day.

Yours very sincerely,

Wilfrid Laurier

Dr. C. O. Fairbank,

Petrolia, Ont.

Letter from Wilfred Laurier

While he was prime minister or leader of the opposition, Laurier wrote several letters to Charles. In this letter, dated October 2, 1911, Laurier writes, "Let me tell you how much I appreciate the great fight you put up in East Lambton." Laurier and Charles had just lost their bid for election in September because the Liberals supported a reciprocal trade agreement with the U.S.

Fairbank Family papers

Leo Ranney

Clara Fairbank had been widowed for two years when she married Leo Ranney, an accomplished mining engineer with Standard Oil. He and Clara's son, Charles, went to Australia together in 1941 to give its government advice on expanding oil production there.

Fairbank Family photo

Chapter 4 – The Third Generation: The Era of Charles Oliver II (Charles Sr.)

Charles Oliver II had a very different upbringing than his father, who had lived with J.H. in their Oil Springs shanty and as an adult worked closely with him. Charles II (who would later be known as Charles Sr.) had moved into the mansion at the age of eight and had a much more limited time with his father, who died when he was just 21.

Like his father, (the doctor, military man and politician), Charles II received a good education. All four sons of Charles and Clara went to Ridley, the private school in St. Catharines. Instead of going to the Royal Military Academy afterward like his father, Charles II obtained a degree in petroleum engineering at the University of California in Berkley.

He was on his way to a career in South America when his mother asked him to return to Petrolia just long enough to put the family's depleting finances in order. He returned to Petrolia with the intent of staying, perhaps a year. This is how it came to pass that Charles became the manager of both Fairbank Oil Properties and VanTuyl and Fairbank Hardware.

When Clara asked her son Charles to return to Petrolia, he did, with the intent of staying perhaps one year.

Once he was back in Petrolia he became involved in the community and became immersed in local politics. From 1934 to 1938, he served as reeve of Petrolia and then became a Liberal member of provincial parliament from 1938 to 1942. He was able to run the oil property and the hardware store at arm's length. While in the provincial legislature, he was able to help local oil producers by introducing a bill that eased the way for land to be leased for oil rights. As well, in 1940, he was able to commission surveys of the historic oil fields of Lambton County. To this day, these documents remain a valuable asset.

While he was an MPP, Charles married Jean Harwood of Moosejaw,

She was an unknown newcomer to town, striding down the main street. "I'm going to marry that girl," he told his friend, even though he had not met her.

Saskatchewan. She had heard of Petrolia when she studied to be a nurse, her instructor, Margaret McPhedran, was from Petrolia. After graduating, she took a nursing job in New York State but was unhappy with its medical practices and she moved to Petrolia to take the position of night supervisor at the Charlotte Eleanor Englehart Hospital.

For Charles, it was love at first sight. Although she was a complete stranger, he saw this striking beauty striding down the main street of Petrolia, then turned to his friend, Dr. MacCallum and said, "I'm going to marry that girl." And he did. Before they had even met, he bought a house with Jean Harwood in mind. Like his parents first did in 1900, he chose a house in the Crescent Park neighbourhood. He bought 425 Warren Avenue.

Their first meeting came when he was feverishly ill in hospital and she nursed him back to health. Once his health was restored, Dr. MacCallum told Charles that Jean had been exceedingly good to him when he was delirious. "You really ought to take her out to dinner," Dr. MacCallum told him. And he did. By all accounts, she had the loveliest legs in all of Petrolia. Some would say her face resembled that of the famed Hollywood actress, Carol Lombard. Others would say she brought Marlene Dietrich to mind. They wed in Moosejaw in 1940. They would have two children, Charles Oliver III (Charlie) in 1941 and Sylvia Jean in 1943.

Charles Sr. chose not to run in the 1943 provincial election. Earlier, he had been elected in a by-election after the death of the sitting member. At the time, he was the youngest MPP in the provincial legislature. After politics, he juggled his time between the two remaining family businesses – VanTuyl Fairbank Hardware in Petrolia and the oil field in Oil Springs. He had fresh ideas for both enterprises. In Oil Springs, he introduced "scale catchers", a type of brass tubing that collected dirt outside the working barrel at the bottom of a well thereby enhanced the well's oil production.

He also worked with Leo Ranney, his mother's second husband. Ranney was credited with developing innovative new technology in horizontal drilling and later adapting it to water wells. During the Depression he created enormous water collector systems that were employed in Lisbon, London and Paris. His water well technology was also employed in American gunpowder plants during World War II.

Together Ranney and Charles Sr. experimented with a type of acid that opened clogged oil wells, and arrived at a technique still used by local oil producers today. In 1941, shortly before Charlie was born, the two travelled to Australia together when the Australian government asked for advice in expanding its oil production.

Like his grandfather, J.H. Fairbank, Charles Sr. dedicated himself to improving life in Petrolia. J.H. was referred to as "the father of the town" and decades later, Charles Senior was affectionately called "Mr. Petrolia". They were instilled with a sense of duty to the community. There was an unspoken belief that looking after your business and your family was not enough; that you owed it to the town to contribute in a meaningful way.

Petrolia had been devastated by the Great Depression and the declining oil industry. Homeless men were allowed to sleep in the basement of Victoria Hall and the residential sidewalks of Petrolia were sometimes marked with mysterious symbols written in chalk. These were messages written in code, saying this family may give you some food to eat.

In Petrolia, as in the rest of Canada, many were without work, money was tight and with no money for upkeep, the homes took on a dilapidated look. With his friend Lew Gleeson and others, Charles Sr. established a sports federation to restore recreation to a demoralized town. After this, he was able to restore Greenwood Park as a space for the Petrolia Fall Fair and sports activities.

At the outset of World War II, it became apparent that the troops needed a drill hall. Charles persuaded the government to restore the derelict buildings of the Petrolia Wagon Works (at the intersection of James and Centre Streets) for this purpose. Another town eyesore was the old Petrolia Imperial Oil refinery (at the intersection of Centre Street and Discovery Line). He took it upon himself to see that the site was bought, cleaned up and used once more. The firm buying the property restored barrels. Vulcan Containers occupies the site today.

There was an unspoken belief that you owed it to the town to contribute in a meaningful way.

Establishing Academic Groundwork

One of Charles Sr.'s most valued accomplishments was to preserve wealth of history that will be prized for generations. He ensured that a veritable mountain of Fairbank papers was put in the capable hands of someone who could tackle the enormous task of organizing them. There are diaries, letters, books, documents, receipts, deeds, accounts, photographs...a century's wealth of writing collected by those who witnessed, lived and made history. Out of dusty diaries, letters in longhand, and piles of yellowed papers, whole stories can be reconstructed - stories of lives, politics, business and events that shook the world.

The job of organizing the Fairbank papers grew to be so monumental, so daunting, that no one even attempted to take it on in all those decades.

Charlie Whipp would play an important role in bringing this goldmine of knowledge to the public. In the early 1950s, Whipp was a reporter with *The Windsor Star,* often going to Petrolia because he knew he could always find a good story there. Twice, Charles Sr. took him to the Fairbank mansion where his mother, Clara Fairbank Ranney, lived.

"He was interested in me because I could get him publicity for the town," said Whipp.

It was on these visits that Whipp first laid eyes on the papers. They were stored on the second floor, in a large room with a turret at one end. The entire room was lined in cupboards and shelves brimming with these papers. "The turret is actually quite big," said Whipp, now semi-retired and living in Kincardine. "It's deceiving from the outside. The Fairbanks once had a boxing ring in there."

Each time Whipp went to the house Charles Sr. said, "One day I'm going to go through all of these papers." At first, Whipp says, he took him at his word. "Then I realized he was never going to do it." There was never enough time and there was never going to be enough time.

After a few years passed, Whipp was reporting for *The London Free Press* and digging up stories at the University of Western Ontario. He told several professors about the Fairbank papers. Among them was Professor James John Talman, who was very receptive to this news because he had done considerable early oil research himself. Talman was the university's chief librarian and a former provincial archivist. "He told me, 'I have a student who I can assign to this. Don't worry about it. Just leave it with me.'"

That young student was Edward Phelps. He was the perfect person for the job; a job that grew to be so monumental, so daunting, no one had even attempted to take it on in all those decades. Phelps had a degree in library sciences from McGill University, a bachelor's degree in English and history, and was working on his Masters degree in the arts, specifically, in Canadian history.

More importantly, Phelps possessed the patience of a paleontologist on a fossil dig, the nose of a bloodhound and a mind like a steel trap. His writings and research have since become legendary in historical circles. He is considered the search and rescue hero of history, famed for "liberating" papers, records and books of all sorts by dashing to the scene moments before they were ingloriously tossed into dumpsters.

"If it weren't for Ted Phelps there wouldn't be any written history of southwestern Ontario," said Bob Taylor-Vaisey, senior records manager of Imperial Oil Archives in Toronto. "He has rescued so much stuff! He is a man with a mission and that mission is to attempt to preserve all records that allows the history of southwestern Ontario to be written."

Phelps and Charles Whipp became a team, writing several historical books together. Whipp too, became an aficionado of Petrolia area history long before he became editor of The Petrolia Topic in 1962.

Initially, Phelps worked in the third floor ballroom of the mansion where there was plenty of room to spread out papers. Additional papers were found above the VanTuyl and Fairbank Hardware store. Still others were found in the basement of the mansion, where termites were munching on three cords of wood stacked there. "The ones at the bottom of these piles were mouldy," said Phelps. "We took them outside to dry on the lawn."

His magnum opus was five years in the making. Then, on May 7, 1965, he presented his professors with 329 neatly typed pages comprising his master thesis, entitled *John Henry Fairbank of Petrolia*, (1831-1914) A *Canadian Entrepeneur.* Though it remains unpublished, copies have been made available for the reference section of libraries.

Thanks to Charles Sr., the public has access to the majority of the Fairbank papers. Phelps says he can't remember if it was his idea to approach Charles Sr. about donating the papers to the university, or Professor Talman's. In any event, many of them were donated to the D.B. Weldon Library at the University of Western Ontario. Papers of a more local interest were donated to

Initially, Phelps worked in the ballroom of the Fairbank mansion where there was plenty of room to spread out papers.

the Lambton Library Headquarters in Wyoming. Phelps, who later became the librarian-in-charge of the Weldon Library's Regional Collection, was instrumental in Charles Sr.'s ability to hand over about 15 nicely organized boxes of material to the Weldon.

Passing down a family business or property to the next generation can lead to the thorny issue of multiple owners.

Charles Sr. went on to help promote oil history in other ways. He later helped The Petrolia Discovery make the leap from the germ of an idea to a reality. In the early 1970s, when a committee was forming to restore the derelict theatre of Victoria Hall, he was a leading light. It was not many years afterwards that director Paul Thompson brought Toronto's Theatre Passe Muraille to Victoria Hall and performed a collective play called Oil, Oil, Oil. The actors had written a play about Petrolia and many of its citizens; Charles Sr. was portrayed on stage by actor Eric Petersen.

Smoothing Out the Tangled Ownership of the Oil Field

It would seem that the family's big, complicated problems had a nasty way of landing on Charles Sr.'s lap. Grappling with the predicament of all the Fairbank papers was a major accomplishment. Another huge challenge he took on was ironing out the tangled ownership of the Fairbank oil property in Oil Springs. His aim was to become the sole owner so that he could make management decisions more easily or sell it. This task took decades, but he persevered and eventually succeeded.

Petrolia had weathered some tough times and the decades had not been kind to Oil Springs either. The period between the 1930s and the 1970s were marked by static prices and climbing costs for oil producers. Most of the oil fields of Ontario were abandoned and Fairbank Oil limped along as best it could. The shared ownership precluded any chance of selling.

All families with businesses or properties that are passed from one generation to another can face the thorny issue of multiple owners. Among the Fairbank family, children would play a pivotal role for they carried on two family legacies, Fairbank Oil Properties in Oil Springs and the VanTuyl and Fairbank Hardware Store in Petrolia.

In simple terms, both businesses have passed from father to son for three generations – J.H. to Dr. Charles Oliver, then to Charles Oliver II (Charles Sr.) and it is now in the hands of Charles

Oliver III, best known as Charlie.

The actual history is much more convoluted. After Edna's death, J.H. lived another 18 years, dying in 1914 at the age of 83. Two years before his death, he let his son Charles takeover the running of his empire. The tangled tale of ownership stemmed from the fact that J.H. had given joint ownership to both of his surviving children: Dr. Charles, then 56, and May, then 45. Upon his death, the oil property was valued at $40,000.

Dr. Charles had managed the business because he lived in Petrolia and because his only surviving sibling could not. May was well ensconced in California and had her family there.

That was fine, but the Fairbank family tree was branching out and shares were divided with each new shoot. Charles and his wife, Clara, were parents to four sons (John Henry II, Charles Oliver II, Henry Churchill, and Robert) and May's marriage to Huron Rock produced three children (John, Margaret, and Warren).

When Dr. Charles died in California in 1925 at the age of 66, his half of the estate was divided among his four sons, giving each a one-eighth share. May, living in California, retained her one-half interest. The divided ownership not only made the oil property difficult to manage it also meant that the meagre profit was shared. Selling was impossible too.

To overcome the divided ownership problems, Charles Sr. began the long process of gathering the shares together. First, he obtained the shares from his three siblings. Buying his youngest brother Robert's share was a straight forward purchase. His older brother, John Henry Jr., had died in 1927, passing his share to his daughter, Claire Fairbank Oakes in England. Claire agreed to sell. In 1962, his younger brother Henry died and that share was willed to Charles Sr. By the mid-1960s, Charles Sr. had successfully reigned in half the ownership.

The other half was then held by May's granddaughter in California, Margaret Rupp, and her descendents. Obtaining these shares was legally complicated because May had stipulated in her will that the land could never be sold. It was also a "per stirpes will"; one dictating that each child's share would be passed to his or her children.

In 1969, Charlie, the son of Charles Sr. and Jean, entered on to this crowded stage. He worked at the oil field for a

The Fairbank family tree was branching out and shares were divided with each new shoot.

In 1973, Charlie bought the property and took over the mortgage.

year and told his father he would like to return at a later date and manage it. Charlie insisted his father should try to overcome this last legal hurdle of ownership.

Finally, in 1973 his father did become sole owner. It was then that Charlie bought the property and took over the mortgage. The site had 350 acres with 70 producing wells and an equal number lying dormant. He incorporated the oil property in 1974 making the official name Charles Fairbank Oil Properties Ltd.

Charles Fairbank Senior

Trained as a petroleum engineer, Charles was on his way to a career in South America when his mother, Clara, asked him to stay and clear up family finances. He intended to stay one year but instead stayed all his life. He became a member of parliament, managed J.H.'s estate, the hardware store and oil property, and his devotion to the town earned him the name Mr. Petrolia.

Photo courtesy of The Hamilton Spectator

Charles Fairbank Senior

Fairbank Oil had been jointly owned by a number of relatives. It took years, but Charles managed to obtain sole ownership.

Photo courtesy of the Hamilton Spectator

Jean Harwood

"I'm going to marry that girl," Charles said to a friend when he first saw Jean Harwood striding down the main street of Petrolia. She was new in town and had come to accept a job as the night supervisor at Charlotte Eleanor Englehart Hospital.

Fairbank Family photo

Greeting Trudeau

A lifelong Liberal and once the youngest member of Parliament, Charles Sr. and wife, Jean, greet Trudeau when he came to Petrolia.

Fairbank Family photo

Unveiling Plaque in Victoria Park

Often called "Mr. Petrolia", Charles Fairbank Sr. is shown here with Gladys Slack unveiling the historical plaque in Victoria Park. The plaque commemorates the founding of Petrolia and was unveiled in June 1974

Photo courtesy of The London Free Press

Magazine Treatment

In the 1950s and 1960s, oil history drew little attention beyond the Oil Heritage District. These two magazines, directed at the oil industry, were exceptions. The 1958 issue of *Canadian Oil News* celebrates 100 years of oil and uses the original trademark of VanTuyl and Fairbank Hardware on its cover. The 1969 issue of *Imperial Oil Review* features a three-pole derrick photo taken at Oil Springs.
- Fairbank Family papers

Charlie Whipp

Charlie Whipp, a newspaper reporter for *The Windsor Star* and later *The London Free Press*, had a key role in writing about Petrolia and Fairbank history. After seeing the vast amount of Fairbank diaries, letters, documents, records, clippings and photos stored in the mansion's turret he told Dr. James Talman at the University of Western Ontario about them. He and Ted Phelps later penned the book, *Petrolia 1866-1966*. Whipp also became the editor of *The Petrolia Topic*.

Photo provided by Charlie Whipp

Edward (Ted) Phelps

Spreading out the Fairbank papers on the floor of the mansion's ballroom, Ted took on the unenviable job of sorting through truckloads of papers. In 1965, he completed his Masters thesis, *John Henry Fairbank of Petrolia (1831 -1914) A Canadian Entrepreneur.* Ted is considered the search and rescue hero of Ontario history, writing 30 volumes on the subject, including *Petrolia 1874-1974*.

Photo by Patricia McGee

Chapter 5 – The Fourth Generation: The Era of Charles Oliver III (Charlie)

No one would have predicted that Charlie would take the helm, not even members of the Fairbank family. There really had been no foreshadowing to his arrival. After all, the family has been based in Petrolia since 1866, not Oil Springs.

He does remember coming to the oil property in Oil Springs as a young boy. But to him, it had simply been the place he and his sister Sylvia chased each other and threw snowballs when the family made its annual expedition to cut down a Christmas tree. As a child, Sylvia, remembers sitting on the jerker line in Oil Springs, riding it back and forth and seeing if she or Charlie could make it stop.

He was certainly never groomed for the role of owner and manager of Fairbank Oil. It didn't appear to be an appealing role for anyone.

Charlie is as surprised as anyone else, for it certainly seemed he was headed in other directions. He took a degree in biology at the University of Western Ontario in London, and later took a job in Toronto, selling textbooks across the province for McGraw Hill publishing company. Afterwards, he spent a year in Montreal, studying history at Concordia University.

In 1969, he chose to go back to Petrolia and work in the Fairbank oil field. He became hooked, but his father warned him that the life of an oil producer was too precarious. Charles Sr. explained in no uncertain terms that his son needed to find another way to earn a living, and if Charlie really wanted to run the oilfield he would at least have something solid to fall back on.

Taking these words to heart, Charlie then obtained a teaching degree from Queen's University in Kingston. For three years he was a secondary school science teacher in Waterford and Pickering. By the fall of 1973, he had his

His father warned that the life of an oil producer was financially too precarious. He needed to find another way to earn a living.

permanent teaching certificate, had fulfilled his promise to his father and was back in the oil field.

Within months, an oil embargo was slapped on the Middle East, the price of oil skyrocketed and headlines screamed of an energy crisis. No one could have possibly predicted that. And Charlie never stood at the front of a classroom again.

Like farmers, oil producers have no control over the price they receive for their product.

Fairbank Oil Grows

Charlie is the first Fairbank since J.H. to sweat in the fields, soak his clothes black with oil and most crucially, devise an arcing vision for the property.

To an oil producer, production is everything. It's the only measurement that counts. Sustaining or increasing the flow of oil is the major preoccupation. Like farmers, they have no control over the price they receive for their product. They are simply told what world prices are on a given day.

Their financial fate largely rests on five factors - what's occurring in the Middle East; what the 11 oil-exporting countries of O.P.E.C. decide at their headquarters in Vienna; oil development around the globe; the actions of politicians everywhere and finally, the vagaries of the stock market.

Several of these factors came into play during two crises for marginal producers in the 1980s – the National Energy Plan of 1980 and the crashed oil prices of 1986. In 1991, the Gulf War erupted and oil prices soared, underlining once again what a crazy and unpredictable business petroleum can be. Also that year, the Ontario government inflicted a new regulation for all oil producers who have always had to dispose the salt water that is produced after the oil is skimmed off. Fairbank Oil was forced to install a new disposal system that carried a price tag of $250,000.

According to the Fairbank papers, J.H. had a sum of 485 wells in Petrolia and Oil Springs yielding more than 25,000 barrels annually in the early 1900s. In 1973, the Oil Springs property produced 10,000 barrels of oil, and by the year 2000 it leaped to 24,000 barrels.

One key reason is that the property has grown. When Charles Sr. had finally become the sole owner and sold it to Charlie in 1974 it had 70 productive wells and 70 idle ones on 350 acres. Today, there are 350 producing wells on 600 acres.

Enlarging the property has provided an economy of scale – more oil flowing with relatively small increases to the overhead operating costs. In 1986, the oil industry was badly shaken when world oil prices collapsed to half their former value. After this financial battering, Charlie felt he needed to devise a survival strategy so that Fairbank Oil could weather whatever oil crashes lay ahead.

When owners with adjacent land offered their properties for sale in the 1990s, Charlie bought them if he thought it made economic sense and the land could feasibly be tied in with existing operations. There were seven purchases within the decade: two properties from Don Matheson in 1991; two properties from David Baldwin in 1992; one from Irv Byers in 1995; one from Paul Morningstar in 1997 and in 1999 he bought from his twin brother, Phil Morningstar. Together, these purchases added up to 250 acres.

It was certainly not a simple matter of taking over the wells on these lands. In some cases they had been abandoned altogether. The Baldwin parcel, for example, had no operating wells and now there are 40. All 400 feet of each well had to be overhauled, cleaned out and put back together. In the case of the Matheson property, a half-century of forest growth made even finding the wells a challenge. Only with the help of a metal detector were some found at all.

After the gruelling job of clearing each well site, bulldozers excavated roads, electric lines were installed and then there was the difficult task of reaming out the wells and rebuilding them. This labour-intensive decade resulted in an additional 200 number of wells for Charles Fairbank Oil Properties.

Often it's assumed that increasing the number of wells has meant a lot of drilling. That is not the case in Oil Springs. Pioneers were so intensive in their drilling in the 1860s that adding new wells today means finding the wells of long ago and making them pump oil again. In the Oil Heritage District of Lambton County men like Charlie are known as marginal producers or strippers. It means that one well produces less than 10 barrels of oil each day.

It was the marginal producers of this area that were financially traumatized in 1980 when the Trudeau government introduced The National Energy Policy. They were lacerated by one prong of the plan; Alberta and other oil producing provinces would be skewered by the plan's royalty regulations.

After this financial battering, Charlie felt he needed to devise a survival strategy so that Fairbank Oil could weather whatever oil crashes lay ahead.

Energy Minister Marc Lalonde looked at the very thin stream of oil, then looked at Charlie. "And what do you do for a living?" he asked.

Spearheading the plan was the Liberal Energy Minister Marc Lalonde.

Boiled down to its essence, the plan said Canadian oil producers would no longer receive the world price. Instead, they would only get the "made in Canada price", a price the government set at 75 per cent of the global price. While the government politely said the National Energy Plan would promote "security, opportunity and fairness", the wounded oil producers screamed "unfair tax grab".

Still infuriated remembering those days, Charlie stated in one of his speeches: "The National Energy Program of 1980 would have put me out of business by 1985...In 1986, (if the plan was fully carried out) the federal government would have been taking five times more revenue than I would, from production which I own and produce."

In the same speech he talked of "the conversion of Saint Marc", an epiphany that happened right in Petrolia. "With the National Energy Plan, Marc Lalonde developed some policies that were not in perfect accord with my interests. I was surprised to find him a charming and engaging man when I met him at The Petrolia Discovery this spring.

"There he saw the largest pumping rig in the world (the Fitzgerald Rig) with a 22-foot bull wheel, gigantic pulleys, enormous belts and a 10-horsepower motor driving jerker lines all over the horizon – all to pump five wells. All very impressive.

"Finally, we saw the result of all this activity - a stream of oil the size of the lead from a pencil – three barrels a day. With a look of incredulity, he turned to me and asked, 'And what do you do for a living?' Three weeks later the government announced marginal well tax relief and effectively absolved us from the Incremental Oil Revenue Tax. Now, I know there was much more going on behind the scene than was evident to me, but I choose to believe that that visit made all the difference, and at that instant, Marc Lalonde was converted."

Oil prices spiked to $40 a barrel at the expectation of the Gulf War of 1991, then settled, then slumped by 1997. At Charles Fairbank Oil Properties, the 1990s were marked by a flurry of activity as new parcels of land were added.

Size had certainly been a major factor in increasing the oil production but it is not the only factor. Looking back to the 1970s, it's evident that increasing production also took an incredible amount of determination and sheer sweat. From the time Charlie arrived in 1973, well up into the 1980s, Charlie put in 14 to 16-hour days in the oilfield. For years, he did his own blacksmithing. Even though he still insists on meticulously making each "packer" that is used in every well, he no longer needs to work alongside his oilmen daily.

Three of today's employees were active participants throughout the 1990s. Dan Whiting earned his first pay cheque from Fairbank Oil in 1979 and was named foreman in the late 1980s. Jaime Collins signed on staff in 1987. Occasionally, generations of the same family have worked at the oilfield. Duncan Barnes has been employed at the oilfield since the late 1980s and his father, Henry, worked during the 1940s and again in the 1970s.

At least three generations of the Sutherland family were foremen. When Charlie arrived, the foreman was Ken Chesney and there were only three other employees: Bucky Mitchell, Henry Barnes and Percy Hawk. Today, the oilfield is usually staffed by five men in the winter. In summer, two to four university students may be hired as well. There is always plenty of work to do, the pump jacks, all 350 of them, operate 24 hours a day, 365 days of the year.

The 1990s were busy in other ways too for Charlie. In 1990, he and Patricia (Pat) McGee became a pair. They first met in the Alberta Rockies in 1978 through their mutual friend, Ron "Cloudy" Beltz, while Pat was working a summer in Lake Louise. Fast forward a dozen years and the three friends gathered together again, this time in London, where Pat was the editor of London Magazine. Charlie and Pat became a couple and Charlie, who loves classical music, had the inspired notion that they marry in the church where Mozart wed in Vienna, Austria. And they did. (While in Vienna, they visited O.P.E.C., the world famous headquarters for the Organization of Petroleum Exporting Countries, but found that guards toting machine guns give very limited access to visitors.) They have two sons, Charles Oliver IV, sometimes called Little Charlie, born in 1992, and Jonathan Alexander, always called Alex, in 1996.

It may appear to some that Charlie's every waking hour is consumed in some aspect of oil but in truth, he is a man of many diverse interests. A partial list includes classical music, foreign films, singing in choirs, astronomy, vegetable gardening, growing trees, live theatre, travel and *Tin Tin* comic books. Unlike

At the O.P.E.C. headquarters in Vienna, gun toting guards give very limited access to visitors.

the past three generations of men in his family, political office of any kind holds no interest for him.

Not all of Charlie's community work centres around oil either. In earlier decades, he has served as treasurer for the Victoria Playhouse Petrolia and was on the executive committee for rebuilding Victoria Hall.

Marginal oil producers have become a rare breed.

Imparting Knowledge of Oil History

It could be said that Charlie's job is to produce oil but his true life's work is imparting his knowledge of the oil history. For this reason, he and his father, Charles Sr., Gary Ingram, Robert Cochrane and a handful of others in the late 1970s developed the idea of The Petrolia Discovery, an outdoor museum with an oil field that works with 19th century technology. It opened in 1980. For more than 20 years Charlie had been volunteering huge chunks of time and enduring every kind of foul weather to maintain its 19th century technology and keep the oil pumping. He could be found there nearly every Saturday afternoon, usually with his friend Cochrane and sometimes with his young sons. In March, 2003, both men retired from Discovery to allow time for other projects. Charlie has also been a valuable resource to the Oil Museum of Canada in Oil Springs ever since he has been in the oilfield.

Always keen to make people more knowledgeable about the oil history, he can be counted on to furnish answers to the media and he is frequently sought after as a guest speaker. Marginal oil producers have become a rare breed and Charlie happily enlightens audiences of many sorts – from the hundreds gathered at the Canadian Petroleum Hall of Fame in Alberta or the Ontario Petroleum Institute's annual meeting to small heritage groups and even elementary schools. His speeches are often remembered and that speaks volumes for their colour, humour and liveliness.

When Gary May was writing his book *Hard Oiler* or when Hope Morritt was writing her book *Rivers of Oil*, Charlie freely gave his time, answering all their questions. Over the years he has appeared on the TV series *Our Town* hosted by Harvey Kirck, and given extensive interviews to *Equinox Magazine*, CBC *Radio*, *The Globe and Mail*, *The Toronto Star*, *The London Free Press*, *The Sarnia Observer, and of course*, *The Petrolia Topic.*

The staff in the oilfield has often aware of these media interviews and Charlie was heartily teased by employee Ken Gould when Charles Fairbank Oil Properties was featured on the *Much Music* segment called *Mike and Mike's Excellent Adventure*. "Wow Charlie! Now you've really hit the big time," Gould said. "It doesn't get any better than this!"

A New Era of Academic Research at Fairbank Oil Properties

But get better it did. It was sheer serendipity that Dr. Emory Kemp of the West Virginia University happened to be driving in the area and saw signs for the Oil Heritage District. He was intrigued and with help from Christopher Andreae from the University of Western Ontario in London, his curiosity soon led him to Charlie Fairbank.

Dr. Kemp is the university's director for the Institute for the History of Technology and Industrial Archaeology and he was utterly fascinated to see the oilfield using technology from the 1800s. He had no idea that this equipment existed. "It was as if they had been studying dinosaur fossils and then came across a living, breathing dinosaur," Charlie told *The London Free Press*.

This was not simply a passing curiosity for Dr. Kemp. He soon organized an American team of university staff and industrial archaeology students and they spent six weeks on the oilfield documenting what they saw. That was the summer of 1999.

Dr. Kemp has been very clear on why this site is interesting to academics. "I believe that the modern petroleum industry can trace its origin to Fairbank Oil and the surrounding oil district. The really important thing is that Fairbank Oil is using mid-19th century technology on a daily basis. There's nothing like that anywhere."

"I believe that the modern petroleum industry can trace its origin to Fairbank Oil and the surrounding district."

The fact that it has been a family operation since 1861 also makes it unique. "I do actually believe it is (UNESCO - United Nations Educational, Scientific and Cultural Organization) World Heritage Site material," said Dr. Kemp. In fact, he has undertaken to make this his personal mission. The documenting is well underway. In the fall of 2002, Dr. Kemp met colleagues in France to discuss a possible World Heritage Site designation.

His enthusiasm after the first field study was infectious and when he returned to Fairbank Oil in October, 2000 it was as the special guest speaker to 70 Americans belonging to the Society for Industrial Archaeology who were getting a detailed, three-day tour of Lambton County's industrial heritage. The tour was organized by Christopher Andreae, Robert Cochrane (a petroleum geologist with Cairnlins Resources Limited and then chairman of Petrolia Discovery), Charlie Fairbank and John Light of Parks Canada. To guide the visitors, Andreae wrote a book, *Lambton's Industrial Heritage*, that is crammed with information. This is important, for it lays down a new layer of documentation that others may build on in the future.

Dr. Kemp's field school at Fairbank Oil included people from The Museum of Science of Technology in Ottawa, Parks Canada, students from West Virignia University, and volunteers who just plainly love industrial archaeology.

In the summer of 2001, Dr. Kemp returned to Fairbank Oil. This time he was leading an official field school. The team members had changed somewhat and included, not only industrial archaeology students and staff, but also employees of Parks Canada (the branch of government that assigns national historic site status), Louise Trottier of The Museum of Science and Technology in Ottawa, and committed volunteers.

They focused on two buildings at Fairbank Oil; the original powerhouse that burned in 1961 and the Orchard Rig. They also studied Albert Baines Machine and Repair Works in Petrolia, which has been making parts for the oil producers since 1914.

The original powerhouse was chosen for study because it is the last remaining remnant of a steam-powered unit on Fairbank Oil. And for proper designation, that's important, said Dr. Kemp. The Orchard Rig was selected because it is typical of the other power rigs on the property and there was not enough time or money to study each one in detail.

Dr. Kemp describes the study as "a three-fold operation". First, the members of the field team measured every inch of these buildings and took the measurements to spacious facilities at The University of Western Ontario where they meticulously drew them to scale. The university became their centre, not only for drawings but for lectures too.

The next step was transferring these drawings to Mylar, a substance ideal for archival work because it is said to last 500 years. It is also well suited to making reproductions. This work was well underway in June 2002. When completed, the originals will be housed in the Library of Congress in Washington, D.C. and copies will be sent to The Museum of Science and Technology in Ottawa.

While in Oil Springs, Dr. Kemp arranged for Hugh Clouse of Clouse Photography to take archival photographs, which are required for official designations. They had to be black and white and on four-by-five-inch negatives. To complete the documenting, Dr. Kemp is writing a report focusing on the refining, pumping and transporting of oil.

One of the highlights of the field school was a speech made by Duane Nellis, the dean of Arts and Sciences at West Virginia University. All the history books were wrong, he told the audience. The Canadians really were the first to start developing oil. The pioneers in Pennsylvania were a close second. This was met with great cheers from all assembled.

Another important aspect of the field school was that officials from Parks Canada were saying that the Oil Heritage District, or portions of it, would likely merit National Historic Site status.

Designations of this kind are not made lightly, nor are they simply bestowed upon a site or area. Instead, they require a great deal of documentation and a group committed to pursuing the application. More work lies ahead. "I do actually believe it is World Heritage material. The problem becomes political," said Dr. Kemp. "I can provide all the technical documentation but politically, it's got to be a Canadian effort."

Charlie has long believed that recognition for the area must begin with the academics. That solid foundation of research is needed before government bodies will open the purse strings to fund interpretive centres. Once the various governments are willing to finance proper facilities, the information will flow through to the public. It has been gratifying for Charlie to see other academics interested in the field in addition to Dr. Kemp and the Society for Industrial Archaeology.

"All the history books were wrong," the American university dean told the audience. "Canadians really were the first to develop oil."

In the fall of 2000, there was another tour of the site, this one for the American Association of Petroleum Geologists. Again, there was exhaustive documentation assembled in the field trip guide book. This study zeroed in on the geological formations and was written by Robert Cochrane.

He has compiled and calculated impressive figures. Using records and carefully calculated estimates, Cochrane finds that the wells of Oil Springs have produced a total 9,930,000 barrels of oil in the period between 1858 and the end of 1999. The Petrolia field produced 18,138,000 barrels in the same time period.

Almost 10 million barrels of oil have been pumped from Oil springs and 18 million from Petrolia.

And he has good news for the Oil Heritage District's future. The amount of recoverable oil in Oil Springs ranges from 250,000 barrels to 1.2 million barrels. This is using current pump jack system. However, Cochrane cautions, the amount that will be extracted in the future totally depends on the price of oil.

'Twas ever thus. In 1866, Oil Springs was abandoned overnight when prices plummeted. It was solely because prices escalated in the 1880s that Oil Springs enjoyed its second boom. Much of the 20th century was an economic lull for local oil producers. It took the energy crisis of the 1970s to galvanize the industry once again with soaring prices.

Being an oil producer has been likened to riding a roller coaster but in truth, a roller coaster is much more predictable. No one knows this better than Charlie Fairbank. And life itself is unpredictable. He points out that the story of Fairbank Oil could have easily turned out quite differently. In two generations, the eldest son died before reaching a 25th birthday, leaving roles open for Dr. Charles and later his son, Charles (Sr.) The legal hurdles erected by Charlie's great aunt May made the land impossible to sell, and his father would have dearly loved to sell it in earlier decades.

"With care and a fair price, more generations will work and learn from the oil field of Oil Springs," says Charlie. "For if, in the earliest days, we exploited the oil, now, we farm it."

Charlie Fairbank

Today, Charlie Fairbank produces about 24,000 barrels of oil each year and is one of the smallest oil producers in Canada. His great grandfather, John Henry Fairbank, produced the same amount in the 1890s and was the biggest oil producer in Canada.

Photo by Hugh Clouse

Blacksmithing

During the 1970's, Charlie studied blacksmithing so that he could make authentic equipment for the oil field. For years he made the hanging irons which supported the jerker line.

Photo by Sylvia Fairbank

Passing On Knowledge

While volunteering at The Petrolia Discovery, Charlie explains the use of an oil can to his young son, Alex. In the background is the impressive bull wheel of the Fitzgerald Rig that measures 23 feet in diameter.

Photo by David Patenaude of The Petrolia Topic

The Current Generation

Unlike his father and grandfather, who lived in Petrolia, Charlie chose to live on the oil property in Oil Springs. He's shown here with his family - his sons Charlie, on right, and Alex, on the left, and wife, Patricia McGee.

Photo by Hugh Clouse

Sylvia Fairbank

Charlie and his sister, Sylvia, grew up in Petrolia. As children, the oil field in Oil Springs was associated with annual expeditions to its woods to cut down a Christmas tree.

Photo by Patricia McGee

Spreading the Word

Grade 5 student Randa Gillatly and Josh Kember of Aberarder Public School penned this letter of thanks after Charlie got the whole class to join their arms and make a "human jerker line". The discovery of oil in Oil Springs, he told them, was a secret.

Fairbank Family papers

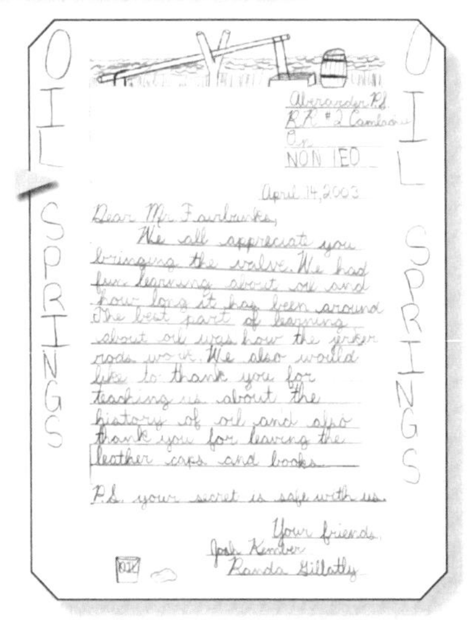

OIL SPRINGS

Aberarder P.S.
R.R. #2 Camlachie
On.
N0N 1E0

April 14, 2003

Dear Mr. Fairbanks,

We all appreciate you bringing the valve. We had fun learning about oil and how long it has been around. The best part of learning about oil was how the jerker rods work. We also would like to thank you for teaching us about the history of oil and also thank you for leaving the leather caps and books.

P.S. your secret is safe with us.

Your friends,
Josh Kember
Randa Gillatly

OIL SPRINGS

Worthy of a Magazine Cover

At Fairbank Oil Properties in Oil Springs, the barn with mural has become a much photgraphed landmark. In 1989, the barn was even featured on the cover of the U.S. magazine *Oil and Gas Investor.*

Fairbank Family papers

A New Era of Academic Research

Dr. Emory Kemp, director of West Virginia University's Institute for the History of Technology and Industrial Archaeology, has been a catalyst for researching Fairbank Oil.

In 1999, he conducted a six-week field study at Fairbank Oil meticulously recording what they saw. He believes Fairbank Oil and the surrounding district is worthy of being a United Nations World Heritage Site. In 2001, Dr. Kemp returned to lead an official field school.

Photo provided by Dr. Kemp

The American Industrial Archaeology Team

The 1999 team was all from West Virginia University, with the exception of Barbara Murray. Seen here in front of the Oil Museum of Canada in Oil Springs are (left to right) Paul Boxley, Barbara Murray, Dr. Kemp, Charlie Fairbank, George Reike, Jeremy Morris and Dan Bonenberger.

Photo provided by Dr. Kemp

Robert Cochrane

For more than 20 years, Charlie and Robert Cochrane worked side by side as volunteers at Petrolia Discovery and both had served as chairman there. A petroleum geologist and engineer with Cairlins Resources Limited, his technical expertise and dedication to historical knowledge have proved invaluable.

Photo provided by Robert Cochrane

PART TWO

Fairbank Oil Properties – Oil Springs

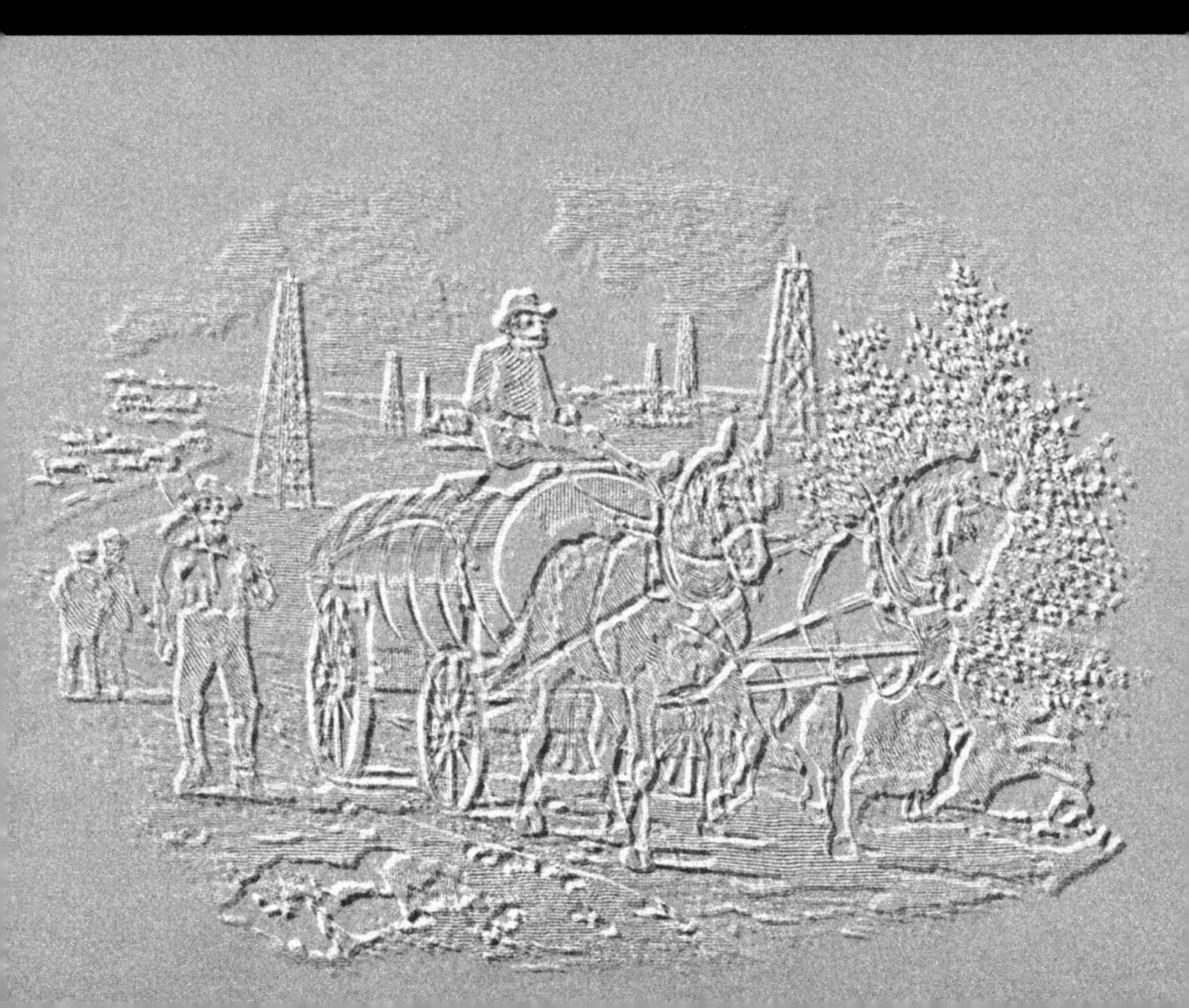

Map of Lambton County, Ontario 2003 showing Rail Lines, Roads and Oil Fields of 1885

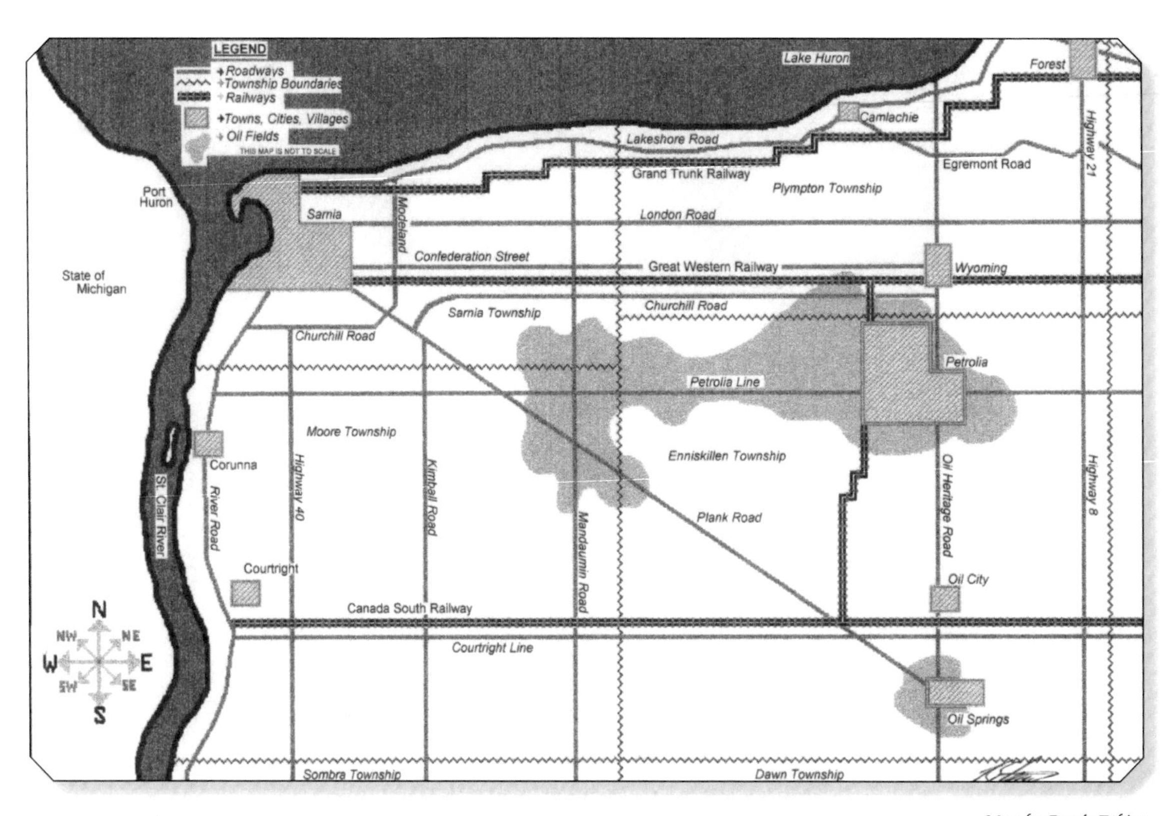

Map by Renée Ethier

Map of the Boundaries of Fairbank Oil Properties

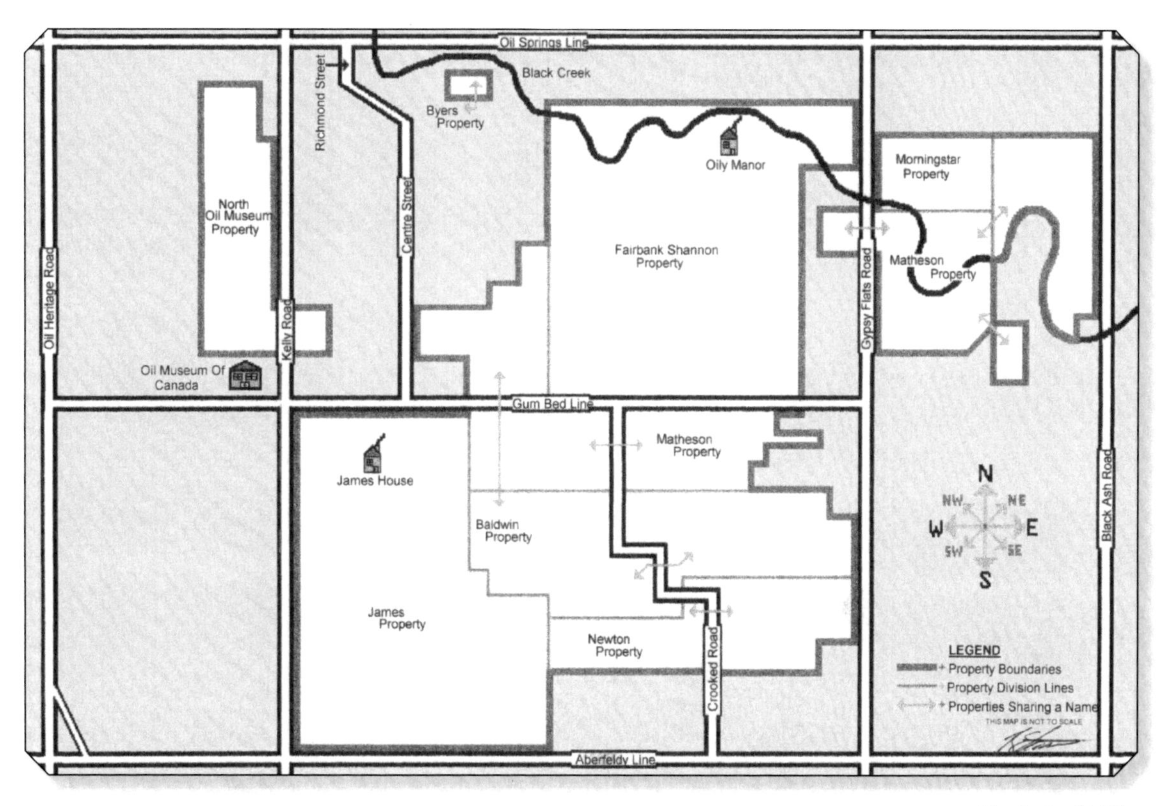

Map by Renée Ethier

Map of the North Section of Fairbank Oil Properties

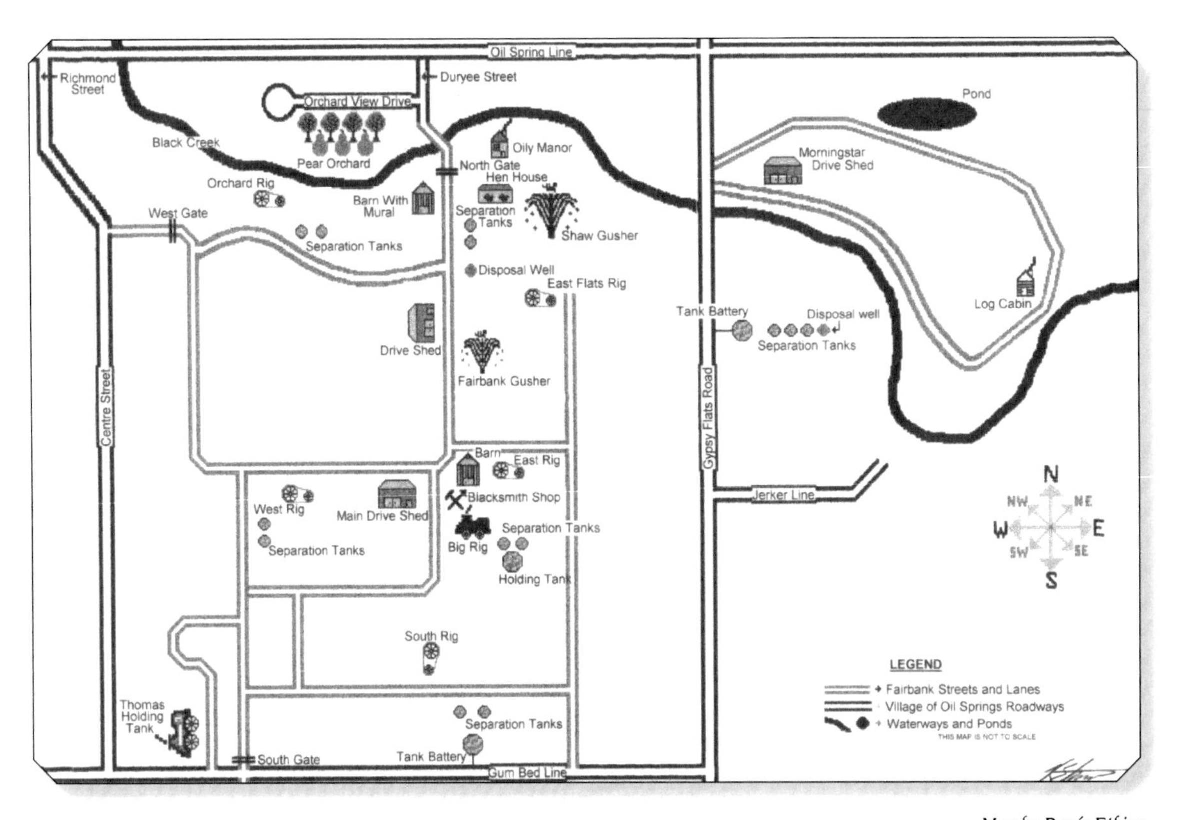

Map by Renée Ethier

Map of the South Section of Fairbank Oil Properties

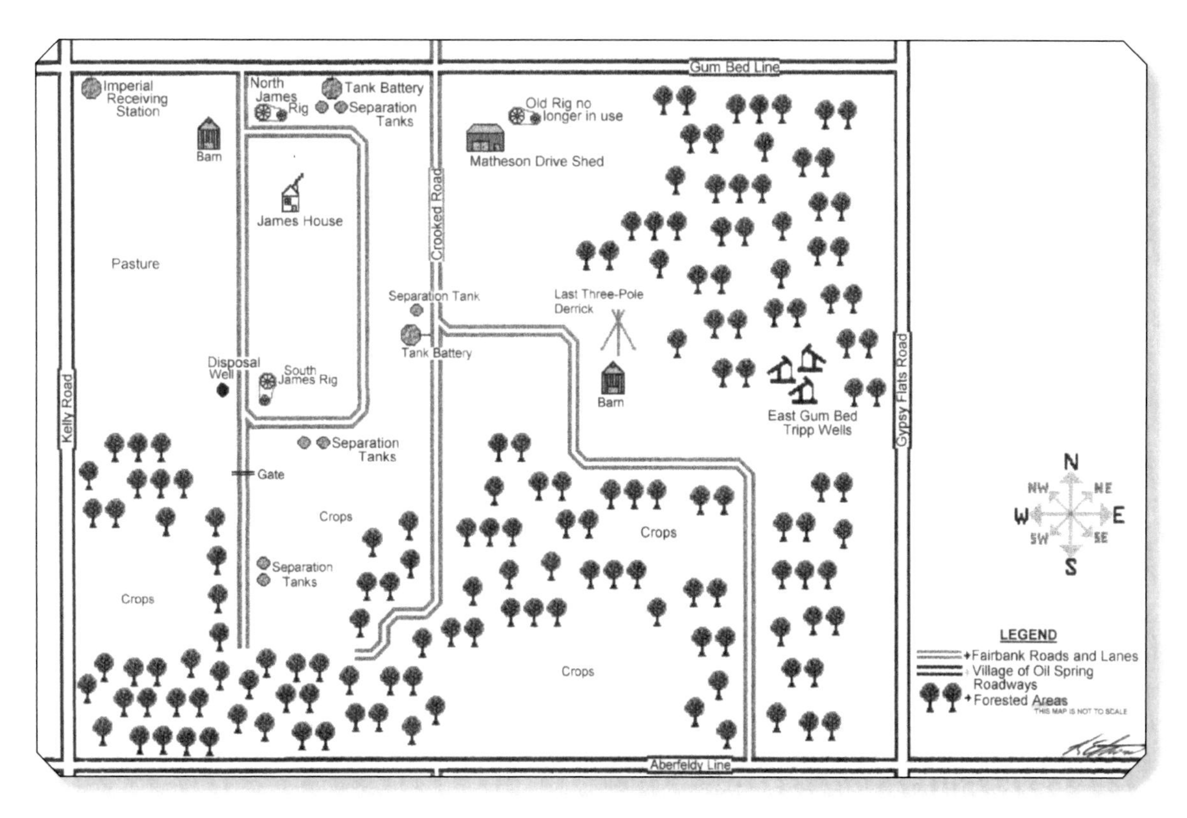

Map by Renée Ethier

Chapter 6 – Historic Sites on Fairbank Oil Properties

The Hand-Dug Wells of the Tripp Brothers – 1850's

Sometime in the mid-1990s, Charlie stumbled upon two hand-dug wells, a number of tanks and the original east gum bed that was developed by Charles and Henry Tripp in the 1850s.

Charles Tripp has been hailed as The Original Oil Man of Canada and together the brothers founded North America's first commercial oil business in 1854 – The International Mining and Manufacturing Company. The site of the wells can be found east of Crooked Road and were discovered only after trees and a good deal of undergrowth, were removed.

The term "hand-dug" is somewhat misleading. It conjures up a visual image of a man deep in a hole, flinging a shovel of dirt 60 feet to the surface. Arnold Thompson of Petrolia explained to Charlie that the pioneers built a tripod over the hole and laid a bridge over the excavation. Dirt was lifted to surface and dumped into a wagon hitched to a horse.

The Site of Shaw's Gusher – Canada's First Oil Gusher, 1862

The famous site where Hugh Nixon Shaw catapulted into the history books by hitting Canada's first gusher in 1862 is on the property. This well, a "rock flowing well", made him instantly rich but was the cause of his death a year later. According to the autopsy, he was asphyxiated by its oil fumes while down his well making repairs.

J.H. Fairbank wrote in his journal February 11, 1863, "Poor Mr. H.N. Shaw drowned in his well today. In him I have lost one of my best friends in Enniskillen. A good man and a most obliging neighbour—sad, sad, sad calamity." A sign on Gypsy Flats Road, south of Oil Springs Line indicates the spot.

This land includes several sites where key moments in oil history happened.

The Black and Matheson "Flowing Well" - 1862

Not quite as famous as Shaw's, but equally historic, are the remains of an 1862 well dug by hand that was the most prolific oil producing well ever tapped in Oil Springs. It was one of the 33 highly prized "flowing wells" of the day and was reportedly producing an incredible 7,500 barrels a day. At a "flowing well" no pump was needed, the oil was under enough pressure that it simply gushed or flowed to surface by itself. All the flowing rock wells and two surface flowing wells had petered out in Oil Springs by 1863.

"I think the Black and Matheson Well produced as much as 6,000 barrels in 24 hours," J.H. told the commission.

"I think the Black and Matheson Well produced as much as 6,000 barrels in 24 hours," J.H. Fairbank told the Royal Commission in 1890. "I made a calculation at the time, and it was about that. That great flow was only of short duration, lasting a few days till it was controlled. Here (meaning Petrolia) the greatest flow was from 400 barrels to 800 barrels, and not much was wasted. These wells continued to flow a good while, and then it became necessary to pump them."

Charlie had always heard stories of this famous Oil Springs well but had no idea where it was located until he spoke to Garnett Byers, who had previously owned oil land with his brother, Bob. They had sold their property to David Baldwin and then Charlie bought it.

"Garney pointed in this direction and said the Black and Matheson Well is over there," Charlie said. Walking through knee-deep grass he noticed the earth formed a low mound - evidence that soil was removed when the well was dug and then dropped beside it, forming a mound. Its location, at the south end of the Baldwin section, matches the vicinity indicated by Garney Byers. "This is the only dug well that I've been able to find in this particular area," said Charlie, "So this is it." Although curious to look inside the well, he feels it's a job best left to archaeologists.

A Three-Pole Derrick - 1860s to 1950s

A three-pole derrick was erected above each well and there had been thousands of them all over Oil Springs and Petrolia. They were usually made from the abundant black ash trees which grew well in the swamps of Enniskillen Township. Now that so much land has been drained, the black ash trees have become rare.

To repair the underground equipment at each well, it first had to be hauled out of the ground. A team of horses supplied the muscle. As the horses pulled ahead, they lifted a cable up through a pulley at the top of the derrick.

Only one three-pole derrick remains on the property, it's on the Crooked Road. It is believed to be the last of the three-pole derricks in Oil Springs and Petrolia that is in its original location. Charlie had been told that it was erected about 1950. "In the early 1950s, a windstorm took down many of our derricks and teamsters were in short supply," said Charlie in one of his speeches. "The teams left our fields then and it is still with some regret, for horses are better than tractors. They do not spin their wheels in the mud. They breed little horses, their fuel is homegrown and they have more character."

The Fairbank Gusher, 1913 – Lambton County's First Gas Gusher

Lambton's first gas gusher was found when Dr. Charles Fairbank was drilling for oil at a depth lower than those that sparked the two previous booms at Oil Springs. When first struck on March 7, 1913, it produced 2 million cubic feet and by March 9, the flow escalated to 11 million cubic feet daily, with 830 pounds of pressure.

Within months, everything changed. The Fairbank Gusher was played out and in May, another gusher was hit 500 feet away and recovered 22 million cubic feet of gas daily. The second gusher occurred when the Oil Springs Oil and Gas Company drilled to 1,906 feet.

The site of both gas gushers can be found slightly northeast of the blacksmith shop.

It's believed to be the last of the three-pole derricks of Oil Springs and Petrolia that is still in its original location.

The Last Imperial Oil Receiving Station to Close, 1974

For decades, Lambton County oil producers relied on the railway to transport their crude to the Imperial Oil refinery in Sarnia. Crude was pooled at receiving stations and then pumped into rail tankers. Later it would be sent by pipeline from the receiving stations. Even after the railway system closed down the receiving stations continued to be used.

Petrolia's receiving station was closed down in 1955. The one in Oil Springs, on land Imperial leased from Fairbank Oil, was closed slightly earlier. Shortly afterward, Charles Fairbank Sr. was able to convince Imperial that the

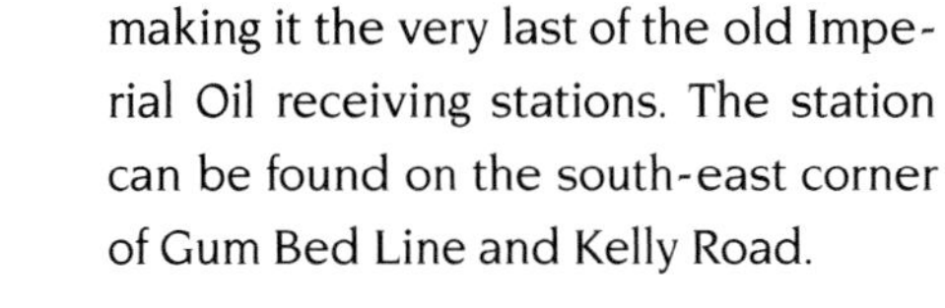

Oil Springs receiving station should remain open. It stayed open until 1974, making it the very last of the old Imperial Oil receiving stations. The station can be found on the south-east corner of Gum Bed Line and Kelly Road.

The site of J.H. Fairbank's original shanty is not on the property.

What is Not to be Found on Fairbank Oil Properties

What's not on this land is J.H. Fairbank's original well "Old Fairbank" nor are there any remains of the 12 by 16-foot log shanty where he lived with his son, Charles, and his mother, Mary, from 1862 to the end of 1865. This land lies west of Fairbank Oil. According to his records, the shanty would have been located close to Centre Street, north and east of the spot where Centre Street intersects with the road leading to the west gate entrance of Fairbank Oil. "Old Fairbank" was close to the shanty.

The parcel of land J.H. was surveying for Mrs. Julia Macklem in 1861, would today be found north of Aberfeldy Road, just east of Gypsy Flats Road.

The Layout of Fairbank Oil Properties

The collected 600 acres of Fairbank Oil is a patchwork quilt made by stitching together several parcels of land. As Fairbank Oil Properties grew in size, it could have become difficult to refer to a particular section. To make communication easier, each of these sections has a name. There was no need to devise new names; each parcel has retained the name of the last owner. Even J.H. used the old names when referring to pieces of his own property.

After all, old names have a way of sticking. Most people in this area do not use the new name Gypsy Flats Rd., instead, it remains The 18th Sideroad. Similarly, The Blind Line is the commonly used name for Gum Bed Line.

When Charlie took the helm in 1973 Fairbank Oil had three main sections – Fairbank & Shannon Property (F. & S.); the James Property; and the Newton Property. Today, the property is described by the following seven parcels of land:

The Fairbank - Shannon Property includes the north entrance from Oil Springs Line and stretches south to Gum Bed Line. This section has the farmhouse, barn, two drive sheds, granary, wood shed, blacksmith shop and five rigs – The Orchard Rig, The East Rig, The West Rig, The East Flats Rig and The South Rig.

The James Property lies south of Gum Bed Line, east of Kelly Road, and includes the James house, the old Imperial Oil receiving station, the North James Pumping Rig.

The Newton Property lies south of Gum Bed Line, and stretches east to border on Gypsy Flats Rd. This property is used for growing soybeans and has oil wells too.

The properties that were added in the 1990s include:

The Baldwin Property is on the west side of the second bend in Crooked Rd. and also north of Gum Bed Line where the Thomas the Tank Engine is placed.

The Morningstar Property is east of Gypsy Flats Rd., between Oil Springs Line and Gum Bed Line.

The Matheson Property is in the woods south and east of Morningstar's. The Matheson 18 Property is south of Gum Bed Line, east of the James Property and north of the section called The South Baldwin Property.

The Byers Property is west of the Fairbank and Shannon entrance.

There was no need to devise new names; each parcel has retained the name of the last owner, even if that man died more than 100 years ago.

Chapter 7 – The Buildings and Sites on Fairbank Oil Properties

The Rigs or Powerhouses and Jerker Line

A pump is needed to extract the oil from a well and the pumps have always required power. Over the decades, these power houses or "rigs" have evolved to meet the needs of the oil property. They are crucial to the oil field. If they are shut down for any reason, the oil wells don't pump and lower production naturally means lower revenue.

Each rig operates around the clock and surprisingly, they only require a 5-horsepower motor. Today, there are six rigs located throughout the 600-acres that comprise Charles Fairbank Oil Properties. (The Orchard Rig, The North James Rig, The South Rig, The East Flats Rig, The East Rig and The West Rig) Together they supply power to 150 of the 350 wells here.

The stories behind these buildings on Fairbank Oil are not documented but in the great oral tradition tales have been passed down through the years and Charlie is continually ferreting out as much information as he can. He has learned that one steam-engine rig was built to power the Fairbank & Shannon Property in 1905 to 1906, and another rig powered The James Section.

Electricity arrived in the oil field about 1918.

Electricity arrived in the oil field about 1918 and at that point the boiler room of the Fairbank & Shannon Rig was converted to a blacksmith shop. Even with one large rig, it was difficult to power more than 200 wells, so some of the wells furthest from the rig were abandoned. To further decrease the workload on the big rig, two extra rigs were built – the Orchard Rig was constructed in 1932 and the South Rig on Gum Bed Line was added shortly afterwards. It's uncertain when the East Flats Rig was built.

Sometime in the early 1930s, the original rig at the James property went up in flames and only the old abutments are still visible along Gum Bed Line. This rig had the job of pumping 65 wells. It was replaced by two rigs-the North James Rig and the South James Rig – and this was considered such an advance in technology that reporters wrote news stories about it. Oil producers rejoiced that wells could be kept pumping even if one rig was shut down for repairs. The North Rig still operates, but the South Rig is now abandoned and replaced by multiple pumping units.

Everything is coated in thick, gooey, glistening black oil. Nothing is blacker.

Fire stuck again in 1961, gutting the big original rig on the Fairbank & Shannon section. Rather than rebuild, two new rigs were constructed, one to the east of it and a second to west. They are known as the East Rig and the West Rig.

Outside, leading into each rig, are the long wooden rods of the double jerker line running parallel to the ground and about eight inches above it. They move slowly (11 times a minute) like two stiff arms pulling back and forth. Each pole has what are called Pitman arms, moving vertically. These arms rest on a set of sliding blocks leading into the rig. Both wooden rods have large turnbuckles that can release the tension when repairs are needed.

Inside the rig feels like a dizzying flashback to the early days of the industrial revolution. It's probably the strong smell of oil that hits the senses first. The clanking and banging is loud. It's very dark. And as the eyes are adjusting there's an awareness that everything is coated in thick, gooey, glistening black oil. Nothing is blacker.

In the middle of the building there is a bull wheel measuring six feet across; at the end of the building, there is a pulley driven by a small motor. A flat belt, measuring 6-inches wide and 36 feet long, almost looking like a gigantic rubber band, is looped around the pulley and the motor. When the bull wheel turns, a small pinion gear drives two large spur gears and they turn two cranks. The cranks act like a bicycle pedal, causing the Pitman arms to move back and forth.

This takes the power outside to a field wheel and shifts the power horizontally. The large cast iron field wheels lie flat slightly above the ground and work like a hub for different "spokes" of jerker lines. The jerker system was designed in straight lines but the wells are scattered about like dandelions. The field wheel was a breakthrough because it linked the different lines of jerker to one power source.

The jerker system, devised by J.H. Fairbank in the 1860s, is often the first thing visitors notice when they come to Fairbank Oil because it looks, well, from another era. "Many people find the jerker line primitive or quaint," Charlie said in one of his early speeches. "Some realize that it is a lot of trouble. One man watching it said it couldn't work. When people ask why I don't modernize and go to pump jacks or secondary recovery, I give them simple answers. Presently, we work 150 wells with four men, two oilers and two pump pullers. To convert to pump jacks would produce no more oil, cost a great deal, save the labour of only one man and triple power consumption....

"The real reason, the one I don't give, is also simple. I don't modernize because it isn't up to me. Oil Springs is a legacy and transcends the individual. It is the creation of all the men who ever worked here. It reminds us of our beginnings. Stripped to the essentials, it teaches men the elements of our business."

The Pump Jacks

In Oil Springs, the oil lies in rock 380 to 400 feet beneath surface and requires an open hole and a pump to get it out of the ground. Metal tubes, called casings, are inserted into the well to keep out shale, water and debris. Two different types of casing are used. The one that runs from surface to the bedrock 30 to 80 feet below is a conductor pipe. It's made of wood. A second, narrower casing fits inside it and runs right down to the oil formation. It's called the production casing.

The pump inside the open hole has lengths of one-and-a-quarter-inch pipe screwed together, reaching a total of 380 feet. At the bottom of the pipe, there is a brass cylinder called a working barrel. Inside it, are two valves. The lower one is stationary. The upper valve is connected to the walking beam of the pump jack by metal rods. The working barrel screws onto a strainer sitting at the bottom of the hole.

Every pump jack has a 10 to 12-foot walking beam that rocks up and down like a teeter totter. The jerker line pulls one end of the beam down, the other end pulls up the rod string and the valves that are in the well. This forces the oil to surface. It then enters a part called the head and is piped to the separating tank.

"Oil Springs is a legacy and it transcends the individual."

The pump jack has become recognized as an international symbol for oil.

Pump jacks at Fairbank Oil receive power in one of three ways. Approximately 200 wooden pump jacks receive power from the rig. The jerker line transfers this power. Another 20 are metal Jones and Hammond Jacks powered by a gearbox. This style of pump jack allows two wells to be pulled in the same direction. Two to six pump jacks can be powered by one gearbox. The third style is a conventional metal nodding horse pump jack and there are about 120 of them on site. These are viewed as temporary because they require an individual motor for power. It's more economical to replace three or four of these with one power source.

The pump jack has become recognized as an international symbol for oil. Whenever one is sighted in Alberta, Texas or anywhere else on the globe it's instantly known you're in oil country. In the mid-1990s Charlie commissioned Sue Whiting to paint dinosaur faces on some of the pump jacks. He claims it was for the amusement of his young sons.

To keep track of the 350 wells on site, they are numbered and also described by their section on the property. Over the years individual wells have occasionally acquired names for a variety of reasons. Any oilman working at the site readily knows where to find The Kenny Rain Dance Well, The Irv Falling Down Well, The Detectorist Well or The Gang of Four.

The Separating Tanks and Disposal System

When the oil is pumped from the ground it's not pure. It's mixed with water. The annual production of 24,000 barrels of oil is usually mixed with about 400,000 barrels of salty water called brine. Each separating tank has what's called a "look box" opening with a lid. Looking inside, the producer can see several pipes pouring oil and water into the tank. Each pipe comes from a specific well so at a glance he can see which of his wells are pumping properly and which are not.

The water has to be separated before the oil can be sent to storage tanks. If it sits long enough, the lighter oil naturally floats to the top of the heavier water. The water is siphoned off and, for more than 130 years, it was fed back into ditches and creeks where an entire eco system thrived on it.

That changed in 1990 when the provincial government deemed it environmentally unfriendly. It declared all brine had to be processed and returned underground within the year. Alarm, anger and exasperation ensued among oil producers. There had been little warning and anyone failing to comply by the deadline would have his operation shut down by the Ministry of Natural Resources. Producers successfully pleaded to have the deadline extended by a year.

The first challenge was to figure out how to do it. Each producer had to design a system that would handle the phenomenal amount of fluid. He then had to build it and make sure it worked. Not going out of business at the same time became an immense challenge. At Fairbank Oil, the disposal system came with a hefty price tag of $250,000. The system is comprised of disposal wells on four sites: one at Fairbank & Shannon, one at East Matheson (acquired when property was purchased 1991), one at James, and another at West Baldwin, drilled in 1995.

The disposal system begins with a six-inch line collecting waste water from an average of 85 wells. The water moves into a tank where it is siphoned into lined tubing taking it 450 feet underground to the porous Detroit River formation in the rock.

Charlie is proud to point out that the design of his disposal system depends only on gravity. Other disposal wells in the Oil Heritage District use electric power to pump the brine. Because Charlie uses no surface pumps to move water, only occasional maintenance is needed. The disposal well can get clogged in the same way a kitchen sink can. Instead of using Drano, as you do in the kitchen, with a disposal well you add a barrel of acid. At Fairbank Oil "going on an acid trip" means you're on your way to picking up a barrel of acid.

At Fairbank Oil, "going on an acid trip" means you're picking up a barrel of acid to unclog a well.

In the first year of the system, Black Creek dried up completely exposing hundreds of clamshells along the creek bed. The water returned to the creek in subsequent years. The change to the local ditches is more pronounced, they are now choked with cattails. The killdeer, once numerous here, have largely disappeared.

The Storage Tanks for Shipping Oil

Once oil is pumped from the ground it's stored until there is a sufficient quantity to be shipped in bulk to Imperial Oil's refinery in Sarnia. The pioneers were painfully aware they were storing a flammable liquid and they feared fire. There were two horrific oil fires. In 1867, virtually everyone would have known that John D. Noble's oil field in Petrolia went up in flames. It was lot 12, concession XI; a mere 100 yards from the majority of Petrolia's biggest wells.

After helplessly watching his oil field destroyed on the night of July 25, Noble wrote, "I looked up and saw that my last tank was gone, and the burning, fiery oil flowed over the land, the flowing well was aflame with all its tanks and the oil hissed and burned and the flames covered a space of ground about a quarter mile wide by a quarter mile long, and leaped 100 feet, and great columns of black smoke rolled up to the sky."

"...the crest of this rushing wave of burning oil gave the creek the appearance of a fiery dragon winding along the valley."

Six men suffered serious burns and the *Petrolia Reporter* stated that J.H. Fairbank, and his party of firefighters, were driven off by the heat the following afternoon. "There was no casing or tubing in the well and the men were at a loss to know how to put the fire out," according to Col. Harkness' manuscript *Makers of Oil History 1850 to 1860*. "Sapping the well" and two wagons of manure finally did the job. "The flow was estimated at 1,000 to 2,000 barrels per day and it had been burning for about one week."

By August 2, another even more horrific fire erupted in the King District. It was the raging Kuwait of its day. It started, according to Harkness, when "the engine man at Lane's well in looking into a 400-barrel oil tank brought his lantern too near the stream of rising gas, which ignited." An astounding 20 acres, equivalent to nine modern football fields, exploded with flames, bursting steam boilers, derricks and flaming debris shooting hundreds of feet into the air.

"Like a present day radio broadcaster he (an eyewitness) tells of workmen connecting a pipe to an oil tank on the border of this inferno in order to remove the oil, that would be further fuel, to a more remote tank, but they only succeeded in spreading the fire to a new area," Harkness wrote.

"The great heat melted steel tanks. These steel or wooden tanks as they caught fire, burst and sent oil along a ditch into a tributary to Bear Creek; the crest of this rushing wave of burning oil gave the creek the appearance of a fiery dragon winding along the valley...At the time he estimated the loss of $100,000 and the fire was still burning."

The two infernos would be a turning point for all oil operations. It led Noble to believe that the safest way to store oil was to deposit it underground. In fact, he poured it right into the clay and claims he never lost a drop.

For the most part, oil has been stored underground ever since. The soil shields against lightning strikes as well as temperature changes that can cause problems in winter when separating out the water. (Separating tanks do a good job of removing the water from the oil but some water remains.)

This method of storage was unique to Petrolia and Oil Springs said Dr. Kemp. It was not used in United States. Of course, the heaviness of the clay in this area is quite unique too. "It's really very ingenious and I think it's entirely safe," he said. Inside these underground tanks was a light cribbing of wood to keep the walls from caving in.

Today, the government insists that oil storage tanks must be above ground; a fact that Dr. Kemp said, "doesn't make any sense to me." On Gum Bed Line one of these tanks can be clearly seen. The tank, inherited from the purchase of the Baldwin property, has never been used here. To the delight of Charlie's two young sons, he hired Renée Ethier to paint the tank to look like Thomas the Tank Engine in 1997.

When the storage tanks at Fairbank Oil are reaching 34 cubic meters (equal to 213 barrels), the foreman calls the Harold Marcus Limited trucking company in Bothwell for pick up. There are five pickup stations here. (Old Stop for the Fairbank & Shannon section, New Stop on the James section, New New Stop on the Baldwin-Matheson section, Stop Four on the Matheson property and Stop 5 on the Morningstar property). Marcus transports all Lambton County oil to the Imperial refinery in Sarnia.

This method of storage was unique to Petrolia and Oil Springs.

The Imperial Oil Receiving Station

The crude in the Oil Springs area used to be collected at an Imperial Oil receiving station. The one-acre site was leased from Fairbank Oil at the southeast corner of Gum Bed Line and Kelly Road, in Oil Springs. "The station in 1933, consisted of four iron tanks and a loading rack 150 feet long with a two-inch pipeline (running) 560 feet long from the station to the loading rack on the Michigan Central Railway siding," according to Don Smith's manuscript *Imperial in the Beginning.*

"The office and pumphouse, 30 feet by 12 feet, of wooden construction, had a five horsepower motor pump for transferring crude oil. Railway cars were loaded and shipped to the 12th

Line Station at Petrolia." (The 12th Line is now called Lasalle Rd.)

Tom Evoy, of Oil Springs, was the receiving station operator, a job previously held by his father. In the early 1950s, Imperial Oil closed the Oil Springs receiving station. Producers then had to make their own arrangements to ship their crude for refining. It's not clear if they had their oil trucked to the Petrolia receiving station or directly to the Imperial refinery in Sarnia.

Closing in 1974, it was the last of all the old Imperial Oil receiving stations. Petrolia's receiving station was the second last to close in 1955.

Shortly after the Oil Springs receiving station closed, a producer from Oil Springs accused Imperial Oil of mixing up the payments. Imperial Oil responded by saying that it "wouldn't purchase any more crude oil from Oil Springs producers unless they each showed proof of clear title to their oil properties," wrote Smith. This posed a further problem for the oil producers.

"Charles Fairbank (Sr.) came to the rescue," Smith wrote. "Charles went to Imperial and proposed that he would purchase the crude oil from the Oil Spring producers and issue them oil receipts, then he would sell the crude oil to Imperial, as the only person Imperial would deal with. Imperial Oil agreed, requesting Fairbank put up a bond."

The agreement lasted about 20 years. After the railway to Oil Springs closed in 1960, the receiving station was still used. Harold Marcus trucked the oil to Sarnia from the receiving station.

When the Oil Springs receiving finally closed in 1974, it was the last of all the old Imperial receiving stations. It marks the end of an era in the history of Imperial Oil and of oil production in Oil Springs. The Petrolia receiving station was the second last to close, and it closed in 1955.

The Oil Museum of Canada has an oil tanker displayed on the exact spot where the railway siding once stood for so many decades.

The Blacksmith Shop

The blacksmith shop was essential at Fairbank Oil and the blacksmith's work was highly specialized. In the pioneer days, virtually everyone did his own blacksmithing. The blacksmith repaired all oil field equipment and forged thousands of "hanging irons" for the jerker system.

Charlie used to regularly blacksmith his own hanging irons after watching and listening to Henry Wheeler at work. Wheeler was a long-time employee, working from about 1914 to the mid-

1970s. Taking a blacksmith course in Detroit in the 1980s, Charlie picked up the finer points of the trade. After tendonitis set in, Charlie chose to use polypropylene rope to replace the iron hangers. It is far more durable than any ropes made in the days of the pioneers. Authenticity is lost by using rope but it is a practical solution.

Traditionally, the blacksmith also supplied horseshoes. Workhorses were used here right up until the 1950s.

Farming, Then and Now

Dates are hazy, but it is known that much of the property was farmed even though it was dotted with oil wells. Crops included corn, hay, oats, wheat, and barley and there were cattle and horses in addition to the sheep and chickens.

Horses were very important to the oil field. They provided basic transportation, hauled oil and were a necessary part of the three-pole derrick system of repairing underground equipment. Horses were bred at Fairbank Oil and at one time there was a horse barn and several other barns. Old horse collars and yokes still hang in the second brown barn, which now serves as a drive shed. It's not known when this barn was raised, likely it was built near the turn of the century. There is also a small, unpainted barn between the two big barns. This has always been known as the granary.

In 2003, about 100 acres are farmed. Seventy-five acres are sown with soybeans and there are approximately 25 of hay grown for the sheep.

For Charlie, it is about six weeks of serious sleep deprivation before all the ewes have had their lambs.

The Sheep and Llamas

Sheep have been raised on the property for 60 years or more. The sheep have three functions. They make great four-legged lawnmowers which reduce the chance of a grass fire, they improve the look of the site and thirdly, they're delicious. They are Suffolk, raised for their meat, not the fine quality of their wool. They are taken to market in Cooksville each fall.

Lambing is always a busy time of year and for Charlie it is about six weeks of serious sleep deprivation before all 100 ewes have had their lambs. The mothers can have gruesome difficulties giving birth or feeding their lambs. It's the bottle feeders that usually cause Charlie the lack of sleep and he has become a knowledgeable midwife.

When the ewes cannot or will not feed their lambs, a milk replacement is

Few realize that the barn mural painting is the original trademark of VanTuyl and Fairbank Hardware.

whipped up in a blender with hot water in the basement of the house, poured into clean beer bottles. If there are extra bottles made, they are refrigerated and later heated in the kitchen's microwave. Newborns need to be fed four times a day. If the lambs are very weak they're put in a box and kept warm in the basement of the house. They sound exactly like newborn babies and in the house there has been occasional confusion between the two.

Each year the sheep also have to be wormed, docked and sheared. Fences need to be maintained and gates must be closed. If they're not, the garden's flowers are munched or the neighbours phone asking if the sheep could please be removed. Rams are kept on site but the challenge is keep them away from the ewes in the fall. If the lambs can be born in the spring, the chance of pneumonia is reduced and they have a better chance of surviving. The large brown barn with the mural is the one used for the sheep.

Since the late-1990s coyotes had been a continual problem. Despite the best efforts of a hired trapper, 20 to 30 lambs were killed by coyotes each year. The answer, Charlie was told, was to get a llama. In the spring of 2003, he bought two – Sam and Gregory. Gregory now patrols Fairbank Oil Properties and Sam is on loan to Ron Brand's flock, which grazes on Fairbank land, both north and south of Gum Bed Line. After several months of their arrival, Charlie happily reported that far fewer lambs have yet been lost to coyotes.

The Barn with the Mural

The much-photographed mural on the north side of the barn is Charlie's brainchild. Inspired by the Joyce Carey novel, *The Horse's Mouth*, he decided he could do what the Gulley Jimson character did in the book - bring a collection of friends together to have a wonderful party and paint a mural.

This was an event. It would be known ever after as the first Lamb Roast. It was the summer of 1981.

The painting depicts a man driving a team of horses that are pulling an oil wagon. Few realize this drawing is the original trademark of the VanTuyl and Fairbank Hardware, the store that opened its doors for business in 1865 and Charlie owns to this day. The image has been heavily copied throughout the Oil Heritage District and a painted version can be seen over the Oil Springs Post Office.

Charlie took the image to his friend and artist Ann Evans, then living in Wyoming, who simplified it and graphed it into one-inch squares. Charlie and friend, Vince Lyons, hammered nails in the barn wall and strung ropes to make one-foot squares. Using the graphed drawing as a guide, Ann and her husband Jim, sketched the outlines of the picture on the barn. The party was in full swing when friends took turns climbing the scaffolding and were directed to "paint their square" in the designated colors.

To top off the festive event, friend Mary Pat Gleeson had arranged for the movie The Horse's Mouth to be shown in the barn. As fate would have it, the wrong film was in the canister but the celebrants, pleased with their artistic contributions, were unperturbed.

Since 1981, more than a dozen not-so-annual Lamb Roasts have been thrown at Fairbank Oil. They bring together a diverse group of friends, family, associates and a great many children for an extended party. The numbers have been swelling and in 2001 dinner was served to more than 400. With the passing of years, the Fairbank Lamb Roast has continually evolved but the barbecuing of the lambs and the sheep-pellet spitting contest (actually chocolate covered raisins) remain sacrosanct traditions.

Until 2003, the barn bore the date 1913 in its foundation. It replaced an earlier barn on the same site that had been destroyed by fire that year. In the summer of 2003, the barn was raised four feet in the air for several weeks while the foundation was ripped out and replaced.

Heated water coils were laid into the cement floor in the area for the ewes and their newborns. Cement floors were added throughout the barn where dirt floors existed before. These and other improvements to the barn have now earned it the name, The Sheep Hilton.

The party was in full swing when friends took turns climbing the scaffolding and were directed to "paint their square" in the colors designated by artist Ann Evans.

The Pear Orchard

Atop the hill on the northwest corner of the property is a pear orchard. Its big soft clouds of white blossoms are carefully watched each year as a harbinger of spring. Usually they bloom shortly before the lilacs and honeysuckle. Sometime in the 1950s, Charlie's father planted about 400 pear trees, not because he was keen on pears but because there was no pension plan for the few oilmen working here. He was particularly concerned about a long-time employee named Bridges and thought this fellow could have some money in his pocket if he looked after an orchard in his retirement.

At the time, there was a pear-canning factory in Petrolia and Charlie's father learned that Bartlett, Anjou and Keifer were the types of pears in demand. He carefully researched how to grow them and then planted. By the time the trees were bearing fruit, these pear types were no longer in demand at the factory.

Ariel Lyons arrived the night before deadline with a children's book of farm animals, confessing she really had no idea how to paint chickens.

Over the years, various friends have been permitted to help themselves to the pears and occasionally the congregation of Christ Church Anglican in Petrolia has harvested the pears as a fundraiser. Shirley Ethier, a good friend of the Fairbank family, usually cans some each year and makes a terrific pear jam. The trees are now left grow wild.

The Chicken Coop

Since there is a chicken coop, it seems safe to assume there must have been chickens raised on the property at one point. There have not been any chickens during Charlie's tenure. The mural was painted by Ariel Lyons, a noted Petrolia artist and former art teacher at the Petrolia secondary school. She's also the owner of Bear Creek Studio in Sarnia, which sells all manner of things, including art supplies.

Charlie commissioned her to paint a mural for the 1982 Lamb Roast and, with the pressure of a deadline, she arrived the night before with a children's book of farm animals, confessing to Charlie that she really had no idea how to paint chickens.

Located at the top of a hill, a two-minute walk from the house, the chicken coop provides handy storage for a diverse collection of toboggans, sleds, flying saucers and rusty old oil equipment. To the Fairbank children, the chicken coop is synonymous with the tobogganing season.

The Entrances

Fairbank Oil Properties has three entrances. The main entrance at the north end of the property is unmistakable. Turning south onto Duryee Street the big barn with the Fairbank Oil, 1861 mural is clearly visible. The gravel road dips down the hill, crosses Black Creek, continues over the jerker line and bends up a hill where 500 daffodils, planted in the fall of 1994, bloom in the grass each spring.

The south entrance on north side of Gum Bed Line, across from Crooked Road, is used as a service entrance. The west gate is accessed from a service road extending from Centre Street. These two secondary entrances become vital when Black Creek floods its banks several days each year making the front entrance impassable.

The House

The wood-frame farmhouse on the hill was built in 1888 according to a foundation block outside the dining room window. This had always been known as the foreman's house and whoever was foreman at the time lived in it with his family. Charlie moved in when Ken Chesney's family moved out in 1980 and he spent the better part of a year renovating it. At that time, he added niceties such as indoor toilets. The outhouse is located east the wooden summer kitchen and is used as a storage shed.

The house is now also home to his wife, Pat McGee, and their sons Charlie and Alex. They've nicknamed their home Oily Manor. It's part of the original Shannon property and was home to three or four generations of the Sutherland family. One branch nicknamed it Paradise, possibly for the great views it affords or because of the many lilac bushes. It's known that fire destroyed the front of the house around 1940.

At one point in the 1970s, Chesney heated the home with natural gas from a gas well located 3,000 feet south of the house. The gas was supplied through a two-inch steel line. "This is an extremely rare occurrence in Lambton County," said petroleum geologist Robert Cochrane. It is common however, in the areas of Leamington, Welland and Niagara.

The pressure on this particular gas well was somewhat low and the line would freeze each time there was a cold snap. Inevitably, it would be the worst days of winter storms when Charlie got phone calls to come and fix it. The Chesney family would then move to Petrolia for a day or two until the lines were thawed and heat was restored. After this happened a few times, Charlie grew exasperated and called the propane company to install a new heating system.

In the 1990s, when installing new water lines to the house, Charlie discovered outer foundations showing the house had once been somewhat larger

Built in 1888, Oily Manor is part of the original Shannon property. It had been "the foreman's house" and home to several generations of Sutherlands.

in earlier days. The house had been painted white, as are so many homes in Oil Springs and Petrolia, but at some point in its long history it had been painted yellow.

Major changes were made to the house in between August, 2002 and June, 2003. An addition almost doubled its size to make it easier to accommodate not only the family, but also visitors attracted to the oil history. The aim was to have modern conveniences without sacrificing the Victorian character of the house. Old windows and doors were replaced, a front porch was added and the exterior wood was removed and replaced with "banana" yellow siding. The architect for the project was John Rutledge, of Goderich, and Dave McCann, of Alvinston, was the head carpenter. During the construction, the brick showing the date of 1888 was partly obscured so it was written again in the addition's stone foundation along with the new date of 2003.

Sometime after the foreman's house was built in 1888, another house was built on the section of land known as the James Property. The date is unknown. It too has been used for employees until recent years and has been rented out since the mid-1990s.

The newest sighting has been the Dunlin, migrating to its summer home in the high Arctic.

The Wildlife

At Fairbank Oil, wildlife flourishes. Though some may think oil production harms the environment, the oil property provides a habitat for grassland, wetland and woodland creatures. In total, there are about 100 acres of woods at Fairbank Oil. These different habitats are quickly disappearing in an age of industrial farming and urbanization.

The animals found here include: deer, ermine, beavers, muskrats, fox, possums, raccoons, groundhogs, coyotes, skunks, bats, rabbits, barn cats, feral cats, black squirrels, frogs, snapping turtles, painted turtles, crayfish, clams, fish, mice, moles and snakes.

Among the birds sighted are: Great Blue Herons, Great Horned Owls, Hairy Woodpeckers, Downy Woodpeckers, Scarlet Tanager, Baltimore Orioles, Cardinals, Turkey Buzzards, Eastern Meadowlarks, Snipe, Killdeer, Barn Swallows, Flickers, Goldfinches, Redpolls, Starlings, Mourning Doves, Chickadees, Blue Jays, different hummingbirds, Robins, Nuthatches, Canada Geese and various hawks, ducks, and sparrows. The newest sighting has been the Dunlin, a red-backed sandpiper that is migrating from its winter home along the coasts of the Southern U.S. to its summer home in the high Arctic.

The Big Rig

This powerhouse, known as The Big Rig, was built in 1906 before electricity arrived. A boiler room next to it supplied the steam and it was able to operate 212 oil wells.

Fairbank Family photo

The Remains of the Big Rig

Henry Wheeler stands beside the remains of The Big Rig after fire destroyed it in 1961. This rig has been studied in archival detail by industrial archaeologists from West Virginia University. Henry Wheeler could have told them a lot. He worked at Fairbank Oil from 1914 to 1970.

Fairbank Family photo

The East Rig

This is one of the six rigs used on Fairbank Oil today to power 350 wells. This particular rig, and another called The West Rig, were built to replace The Big Rig after the fire. The bull wheel on the left is connected to the gears by a common shaft.

Fairbank Family photo

The Remains of the Boiler Room

After electricity arrived around 1918, the boiler room was converted to a blacksmith shop.

Photo by Willy Waterton

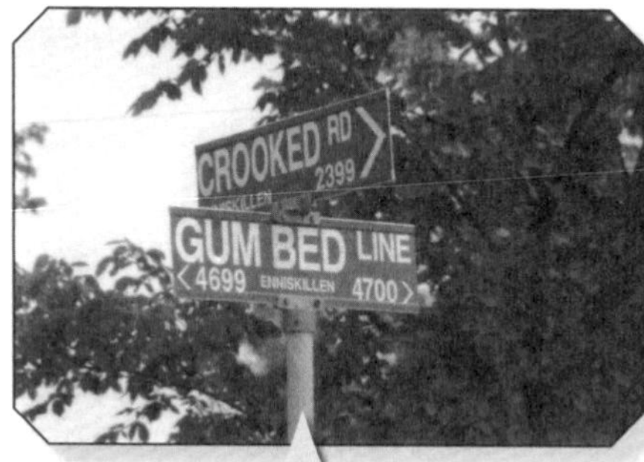

Getting Your Bearings

Fairbank Oil is a 600-acre patchwork quilt made up of several parcels of land. Parts of it are north and south of Gum Bed Line, other parts are east and west of the Crooked Road.

Photo by Willy Waterton

The Field Wheel

The cast iron field wheel, also known as the spider wheel, acts like a hub for several jerker lines radiating out in different directions. Field wheels are usually located close to the rig and the jerker line sends power from the rig to the wells.

Photo by Willy Waterton

The Pump Jack

Even the ice storm in 2003 didn't stop the pump jacks from bringing the oil to surface. As the wooden walking beam nods up and down above ground it is lifting and lowering a valve assembly 400 feet below. This pumps the oil to surface.

Photo by Patricia McGee

The Separator Tank

Today, for each gallon of oil pumped to surface, there is about 20 gallons of water pumped with it. In the separator tank, oil floats to the surface and the water below it is siphoned off to a disposal well.

Photo by Willy Waterton

The Jerker Line

Animals have no fear of the jerker line. The sheep leap over it easily and Gregory, the llama, is quite at home with it. The llama can be seen behind the pump jack in this photo.

Photo by Patricia McGee

The Jerker Line and Pump Jack

The jerker line and pump jacks run 24 hours a day, every day of the year.

Photo by Patricia McGee

A Former Oil Crew

This shows the oil crew in 1969. From left to right are: Ken Chesney; Frank Selman; George Howlett; Henry Wheeler; Reg Sutherland (foreman at the time); Ron Sutherland; Irv Henderson, and Charlie.

Photo by John Hus, The Sarnia Observer

The Last Three-Pole Derrick

Once there were thousands of three-pole derricks in Oil Springs and Petrolia. Charlie stands beside the last one to remain in its original location. The derrick had a pulley at the top which allowed the rods and tubing to be hoisted out of the ground for repairs. Horses supplied the power and these were used until the 1950s.

Photo by Patricia McGee

The Portable Pulling Rig

Instead of erecting a three-pole derrick above each well, it saved labour to make the derrick portable. Horses pulled these well into the 1950s in the Oil Heritage District.

Photo by John Hus, The Sarnia Observer

Connecting Pipes

Foreman Dan Whiting uses Monkey Grip gloves to screw pipes together.

Photo by Willy Waterton

Running The Pulling Rig

Foreign driller Sam Donaldson, seated, and Ted Morningstar, an Oil Springs producer, sand pumps a well at Fairbank Oil in 1969.

Photo by John Hus, The Sarnia Observer

Cleaning The Sand Pump

Dan Whiting removes the cup of the sand pump which takes out debris from the well.

Photo by Willy Waterton

Pulling A Sand Pump

Duncan Barnes hoists a sand pump. He's known oil all his life and even as a child he worked with his father, Henry, who earlier looked after the James property at Fairbank Oil.

Photo by Willy Waterton

Checking The Tank

Loyd Woods, an Oil Springs producer, checks his separating tank. Each pipe feeding into it is connected to an oil well so a producer can see how much each well is producing.

Photo by Willy Waterton

Removing Pipe

Dan Whiting, left, and Jaime Collins, right, unscrewing pipe as they clean out a well.

Photo by Willy Waterton

New Foundations for an Old Barn

In the summer of 2003, the 1913 barn was lifted four feet in the air, then supported for weeks by steel H-beams. The old foundation was ripped out and replaced. Charlie is seen here with his two sons, Charlie, on left, and Alex on the right.

Photo by Phil Hein

Thomas the Tank Engine

To amuse his young sons, Charlie commissioned Renée Ethier to paint an empty oil tank as Thomas the Tank Engine. It is visible on Gum Bed Line.

Photo by Patricia McGee

1861
FAIRBANK OIL

Chapter 8 – Epilogue

Fairbank Oil Properties is more than just a long-standing oil business. History that would reverberate around the globe was made right on this site. Within its 600 acres are the earliest works of Charles Tripp, Canada's original oil man, who found the gum beds and founded North America's first commercial oil well.

Other historical sites can be found at Fairbank Oil too. The site of Canada's first oil gusher, drilled by "that crazy Yankee" Hugh Nixon Shaw in 1862 is also here. One of the original "flowing wells", the Black and Matheson well of 1862, has been located as well. Canada's first gas gusher, The Fairbank Gusher of 1913, and its second gas gusher are here. And there remains standing but one of the hundreds original three-pole derricks that once dotted the landscape.

The business, now more than 142 years old, has passed down three generations of Fairbank men. Many ask Charlie if his two sons will carry on the legacy. As of this writing in 2003 the oldest boy, Charlie is only 11 years old. Alex is a mere seven years old.

All a father can say is that it will be up to them to choose. All a father can wish for is that they each follow their dreams. What dreams those may be, no one knows.

What dreams those may be, no one knows.

PART THREE
The Times of Fairbank Oil and the Fairbank Family

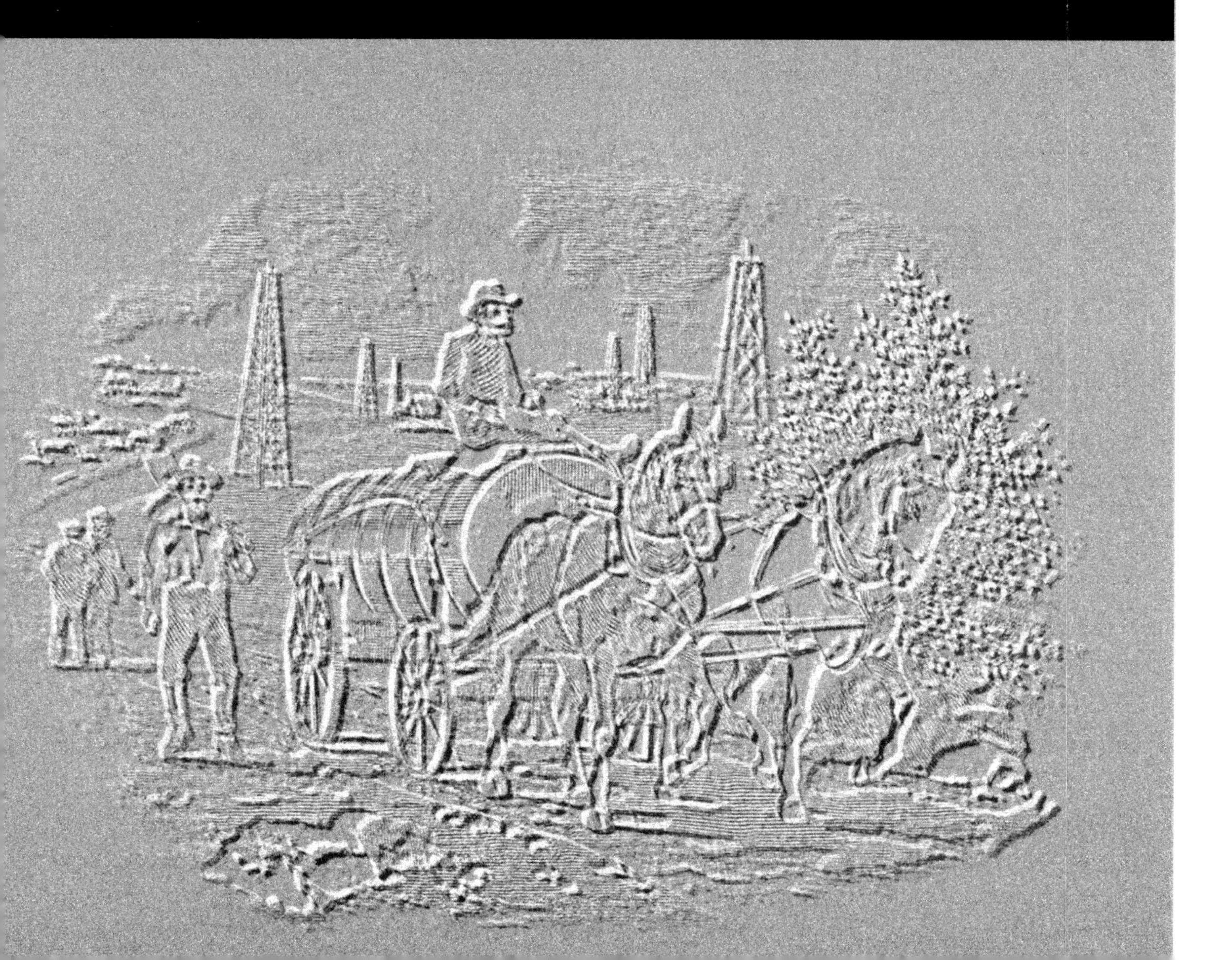

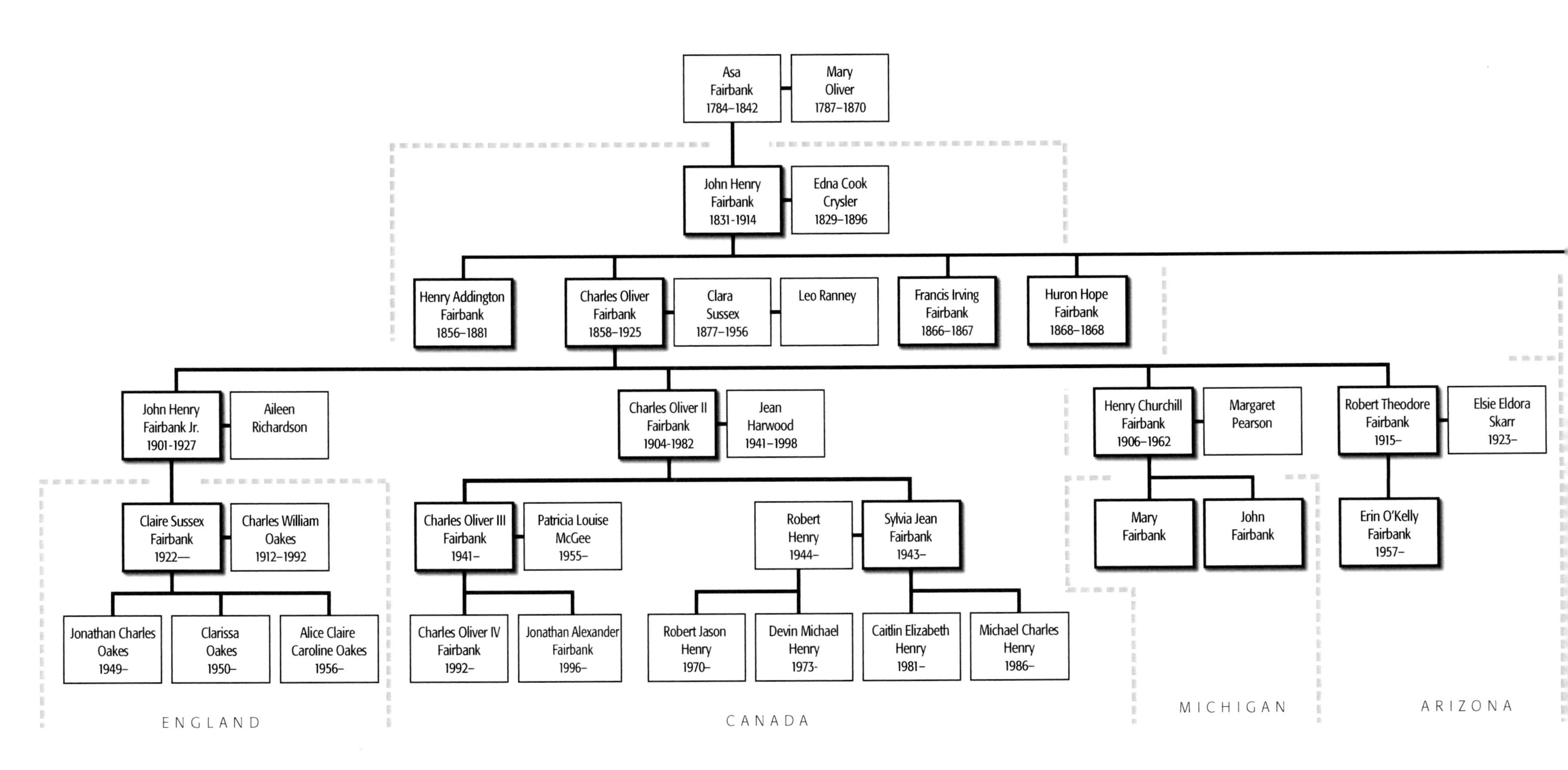
Asa Fairbank 1784–1842
Mary Oliver 1787–1870
John Henry Fairbank 1831-1914
Edna Cook Crysler 1829–1896
Henry Addington Fairbank 1856–1881
Charles Oliver Fairbank 1858–1925
Clara Sussex 1877–1956
Leo Ranney
Francis Irving Fairbank 1866–1867
Huron Hope Fairbank 1868–1868
John Henry Fairbank Jr. 1901-1927
Aileen Richardson
Charles Oliver II Fairbank 1904-1982
Jean Harwood 1941–1998
Henry Churchill Fairbank 1906–1962
Margaret Pearson
Robert Theodore Fairbank 1915–
Elsie Eldora Skarr 1923–
Claire Sussex Fairbank 1922—
Charles William Oakes 1912–1992
Charles Oliver III Fairbank 1941–
Patricia Louise McGee 1955–
Robert Henry 1944–
Sylvia Jean Fairbank 1943–
Mary Fairbank
John Fairbank
Erin O'Kelly Fairbank 1957–
Jonathan Charles Oakes 1949–
Clarissa Oakes 1950–
Alice Claire Caroline Oakes 1956–
Charles Oliver IV Fairbank 1992–
Jonathan Alexander Fairbank 1996–
Robert Jason Henry 1970–
Devin Michael Henry 1973-
Caitlin Elizabeth Henry 1981–
Michael Charles Henry 1986–
ENGLAND
CANADA
MICHIGAN
ARIZONA

Chart 1
Fairbank Family Tree

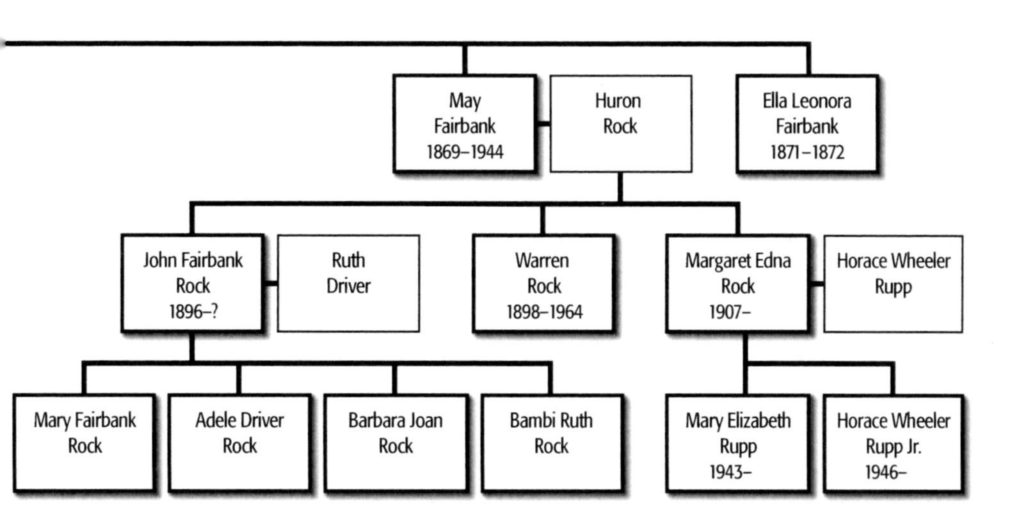

Chapter 9 – Timeline

1784	Sept. 9	~ **Asa Fairbank**, father of John Henry, born in Athol, Massachusetts
1787	Feb. 25	~ **Mary Oliver**, mother of John Henry, born in Athol, Massachusetts
1820		
1829		~ Asa Fairbank marries **Mary Oliver**
	May 11	~ **Edna Crysler**, (future wife of John Henry Fairbank) born in Niagara Falls, Ont.
1830		
1831	July 21	~ **John Henry (J.H.) Fairbank** is born in Rouse's Point, New York (only child)
1836		~ Canada East gets its first railway
1837		~ Queen Victoria begins her reign
1840		
1840s		~ The telegraph arrives in Canada East and Canada West
1842	Oct. 26	~ **Asa Fairbank** dies at age 58. J.H. is 11 years old at the time
1850		
1852		~ Charles Tripp petitions to get his Oil Springs firm, International Mining & Manufacturing Company, incorporated for commercially producing oil products
1853		~ J.H. arrives in Niagara Falls, Ontario, Canada
1854		~ Abraham Gesner of Nova Scotia gets U.S. patent for kerosene made from coal
	Dec. 18	~ Charles Tripp gets approval to found North America's first commercial oil company

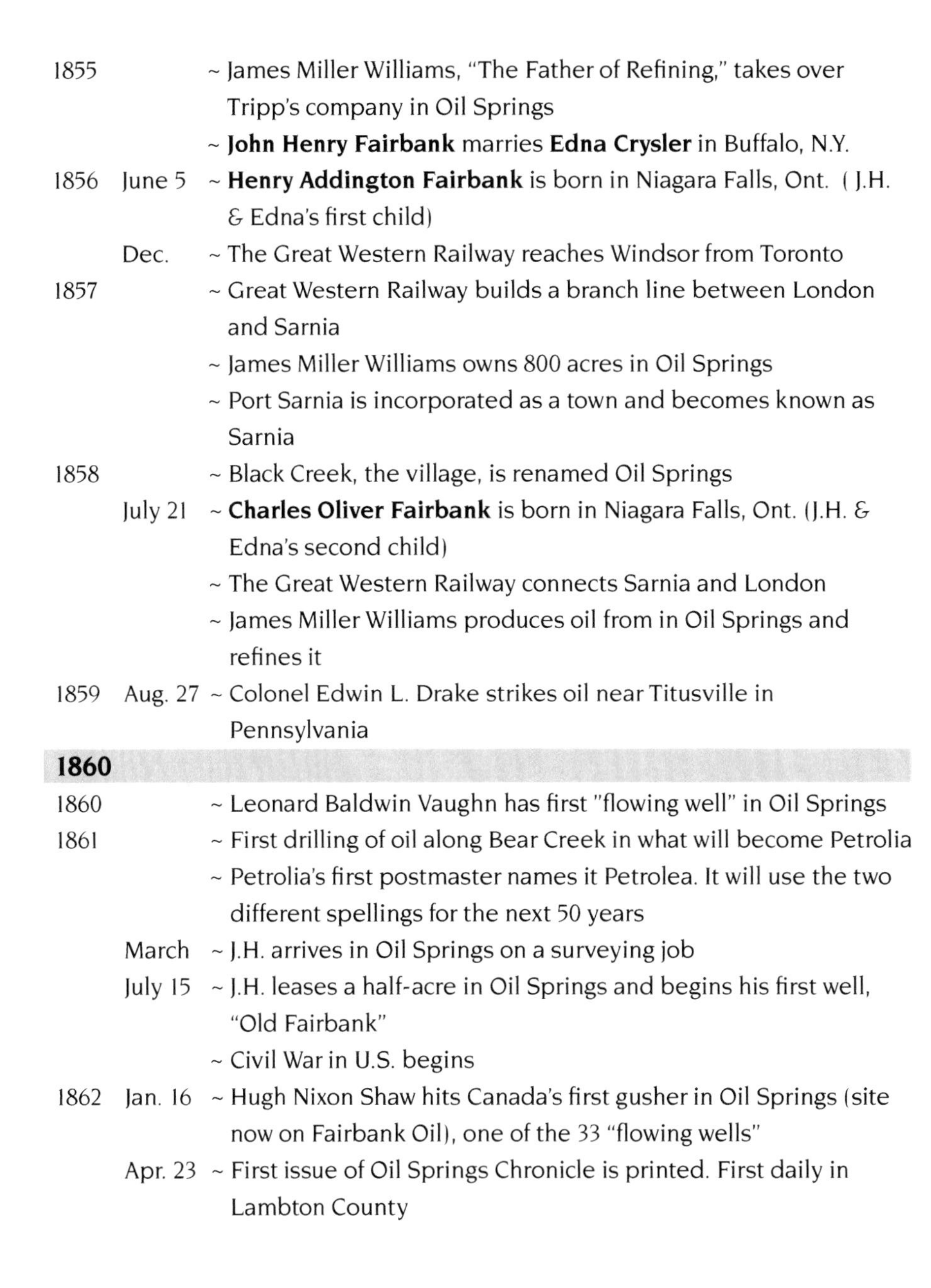

1855		~ James Miller Williams, "The Father of Refining," takes over Tripp's company in Oil Springs
		~ **John Henry Fairbank** marries **Edna Crysler** in Buffalo, N.Y.
1856	June 5	~ **Henry Addington Fairbank** is born in Niagara Falls, Ont. (J.H. & Edna's first child)
	Dec.	~ The Great Western Railway reaches Windsor from Toronto
1857		~ Great Western Railway builds a branch line between London and Sarnia
		~ James Miller Williams owns 800 acres in Oil Springs
		~ Port Sarnia is incorporated as a town and becomes known as Sarnia
1858		~ Black Creek, the village, is renamed Oil Springs
	July 21	~ **Charles Oliver Fairbank** is born in Niagara Falls, Ont. (J.H. & Edna's second child)
		~ The Great Western Railway connects Sarnia and London
		~ James Miller Williams produces oil from in Oil Springs and refines it
1859	Aug. 27	~ Colonel Edwin L. Drake strikes oil near Titusville in Pennsylvania
1860		
1860		~ Leonard Baldwin Vaughn has first "flowing well" in Oil Springs
1861		~ First drilling of oil along Bear Creek in what will become Petrolia
		~ Petrolia's first postmaster names it Petrolea. It will use the two different spellings for the next 50 years
	March	~ J.H. arrives in Oil Springs on a surveying job
	July 15	~ J.H. leases a half-acre in Oil Springs and begins his first well, "Old Fairbank"
		~ Civil War in U.S. begins
1862	Jan. 16	~ Hugh Nixon Shaw hits Canada's first gusher in Oil Springs (site now on Fairbank Oil), one of the 33 "flowing wells"
	Apr. 23	~ First issue of Oil Springs Chronicle is printed. First daily in Lambton County

		~ Charles Oliver, 4, and Mary Oliver Fairbank, 75, arrive in Oil Springs to live with J.H.
		~ J.H. is sued for oil debts four times
		~ J.H. records his best day yet, profits $150 from "Old Fairbank" well
		~ Jerker line is devised by J.H., allowing many wells to be pumped with one steam engine
		~ Black & Matheson prolific "flowing well" is struck in Oil Springs (site now on Fairbank Oil)
		~ J.M. Lick discovers oil in Bothwell
		~ Plank Road is built from Oil Springs to railhead in Wyoming
1863		~ "Flowing wells" finish in Oil Springs but oil is still abundant
	Dec. 16	~ Oil Springs is incorporated as a village
1865		~ J.H. sells "Old Fairbank" well for a handsome profit
		~ J.H. starts buying land in Petrolia
		~ J.H. opens a store in Petrolia which will later be known as VanTuyl and Fairbank Hardware
	Dec.	~ J.H. moves into his wood frame house in Petrolia with his mother and his son, Charles
		~ Civil War ends in the U.S.
1866	May 1	~ Edna and 9-year old son, Henry, move to Petrolia, family reunites
	June 1	~ American Fenians defeat Canadian militia at Ridgeway in Niagara
		~ Construction of a 108-bedroom hotel completed in Oil Springs but never opens
	June 8	~ Fenians attack Quebec
	Nov. 23	~ Captain B. King hits major oil find in Petrolia and kicks off town's 40-year oil boom
	Dec. 25	~ **Francis Irving Fairbank** is born in Petrolia (J.H. and Edna's third child)
		~ Petrolia incorporated as a town

1867	Jan. 1	~ Petrolia citizens finance their own five-mile rail spur from Petrolia to Wyoming
	July 1	~ Confederation of Canada
	July 25	~ John D. Noble's Petrolia oil field explodes in flames. Six badly burned
	Aug. 2	~ A horrific 20-acre (9 football fields) oil fire in the King Well area
	Aug. 9	~ **Francis Irving Fairbank** dies at 7 months
		~ John D. Noble devises underground storage tanks
1868	June 2	~ **Huron Hope Fairbank** born in Petrolia (J.H. & Edna's fourth child)
	Aug. 9	~ **Huron Hope Fairbank** dies at 2 months
1869	Aug. 10	~ Vaughn & Fairbank, Bankers opens in Petrolia (known as The Little Red Bank)
	Oct. 25	~ **Mary Edna (May) Fairbank** born in Petrolia (J.H. & Edna's fifth child)

1870

1870's		~ Canada's first oil pipelines begin in Petrolia
1870	June 8	~ **Mary Oliver Fairbank**, J.H.'s mother, dies in Petrolia at age 83
1871	Dec. 6	~ **Ella Leonora Fairbank** is born in Petrolia (J.H. & Edna's sixth child)
1872	Aug. 30	~ **Ella Leonora Fairbank** dies at 8 months
1873		~ J.H. and others form Home Oil Company, J.H. serves as president and general manager
1874		~ Queen Victoria's Diamond Jubilee
		~ J.H. starts his first of three terms as chief of the Petrolia Fire Brigade
	July 1	~ Major Benjamin VanTuyl becomes a partner with J.H. in the hardware store
1875		~ Vanalstyne and Smith build a two-mile pipeline from Marthaville to their tanks
1876		~ Charles Fairbank (J.H & Edna's second child) enters Royal Military College in Kingston, Ont.

1877 July 17 ~ **Clara Sussex** (future wife of Charles) born in Vanessa, Ont. (raised in Bothwell)

Oct. 24 ~ Mutual Oil Association is formed in a first attempt in Petrolia to unite oil producers and regulate prices

1879 ~ Jake Englehart builds Silver Star in Petrolia, the largest refinery in the country

May 1 ~ Black Friday for oil producers, oil plummets to 40 cents a barrel

~ Mutual Oil Association collapses

1880

1880's ~ Horse-drawn wagon transport peaks, age of pipelines develops

1880 April 30 ~ Imperial Oil is formed in London, Ont. by 19 refiners

Spring ~ Charles Fairbank is in first graduating class of Royal Military College

~ Charles takes commission in England of Royal Artillery, attends Woolwich Academy

Aug. 12 ~ Contract shows J.H. Fairbank selling crude to Imperial Oil (Proof Fairbank Oil has been supplying crude to Imperial Oil longer than anyone)

Sept. 8 ~ Imperial Oil is incorporated

1881 Jan. ~ J.H. sells Home Oil Refinery to Imperial Oil

Feb. 1 ~ **Henry Addington Fairbank** dies in Ann Arbor, Michigan at age 24 (Charles and May become J.H. & Edna's only surviving children)

~ Charles returns from England after death of older brother and later takes commission of London Artillery Field Battery

~ W.S. Duggan of Excelsior Oil Company finds deeper oil in Oil Springs and kicks off second boom

1882 ~ Imperial Oil moves its barrel plant to Petrolia

~ Canada Southern Railway extends to Oil City on St. Thomas-Courtright line

~ Pipeline is built from Oil Springs to Petrolia by Petrolia Crude and Tanking Company

June 20 ~ J.H. elected as Liberal MP for East Lambton

		~ J.H. opens Crown Savings and Loan in Petrolia and is president for next 30 years
		~ J.H. buys two-thirds interest in Shannon oil property in Oil Springs
1883		~ A second pipeline is built from Oil Springs to Petrolia by the Crown Warehouse Company
		~ Imperial Oil's refinery in London is destroyed by lightning fire
1884	Dec.	~ Imperial Oil moves its head office from London to Centre Street in Petrolia
	Dec. 23	~ Petrolia Oil Exchange opens, J.H. will remain president until 1897
1885	Nov. 7	~ Last spike of Canadian Pacific Railway is hammered in at Craigellachie, B.C., giving Canadians rail from the Pacific to the Atlantic Ocean
1886		~ Spur line of Canada Southern railway reaches Oil Springs from Oil City
1887		~ J.H. defeated in federal election
		~ Charles takes command of the 6th London Field Battery
1888		~ White farmhouse, "foreman's house" built on Shannon property at what becomes Fairbank Oil (now home to Charlie Fairbank & family)
		~ Charles leaves to study medicine at Columbia University in New York City
1889		~ Construction of Fairbank mansion, Sunnyside, begins
		~ Victoria Hall opens in Petrolia
		~ J.H. steps down as chief of the Petrolia Fire Brigade

1890

1890	Sept.	~ J.H. buys remaining one-third share of Oil Springs oil property from Shannon
		~ Report of the Royal Commission on the Mineral Resources of Ontario is published
1890s -1900		~ J.H. buys the James Property in Oil Springs, bringing his property up to 160 acres

		~ James House on Gum Bed Line is constructed on Fairbank Oil (date approximate)
		~ J.H. is largest oil producer in Canada (his production includes Petrolia and Oil Springs)
1891		~ William Stevenson, owner of Stevenson Boiler Works, Petrolia's biggest non-oil industry, slinks out of town owing $14,000 to The Little Red Bank. J.H. buys it at sheriff's auction, refits the plant and runs it until 1918
		~ Fairbank mansion is completed, was the largest private home in Lambton County
		~ Charles receives medical degree from Columbia University in New York
1894		~ Edna retires in Pasadena, California with daughter, May
		~ Charles finishes post-graduate work at Columbia University
1896	Jan. 1	~ **Mary Edna (May) Fairbank** marries **Huron Rock** in Pasadena, California
	March 7	~ **Edna Fairbank** dies in Pasadena, California at age 66
	Dec. 5	~ **John Fairbank Rock** born in Pasadena, California (first child of May and Huron. Later children are Margaret, then Warren)
		~ Bushnell Oil Co. buys Fairbank, Rogers & Company Refinery plus Queen City Oil in Petrolia
1897		~ Bushnell Oil Co. buys Alpha Oil Refinery in Sarnia
		~ Imperial Oil moves head office from Petrolia to Sarnia, then its refinery
1898	July 1	~ Standard Oil buys 75 per cent of Imperial Oil
		~ Charles Fairbank and Frank Carmen find new oil in Bothwell, starting a second boom
	Sept .	~ J.H. lays one of the first pieces of concrete sidewalk in Petrolia, in front of the Vaughn Block
1899		~ Britain declares Boer War. Charles recruits troops. Total of 6,000 Canadians enlist

1900

1900	July 5	~ **Dr. Charles Oliver Fairbank** marries **Clara Sussex**
1901	April 11	~ **John Henry Fairbank II** is born in Santa Barbara, California (Dr. Charles & Clara's first of four sons)
		~ Queen Victoria dies. King Edward VII reigns
		~ J.H. donates main street lots to Town of Petrolia for what is now Victoria Park
	Aug. 30	~ Western Canada's first producing oil well in Waterton National Park
1902	May 31	~ Britain declares victory in Boer War
1903		~ John Noble's company, Canadian Oil Fields Limited, builds the world's largest pumping rig in Petrolia, The Fitzgerald Rig, capable of pumping 350 wells at once
1904		~ J.H. improves the fire extinguisher manufactured at Stevenson Boiler Works, the Fairbank Fire Extinguisher sells well
	June 6	~ **Charles Oliver Fairbank II** ("Charles Senior") is born in Petrolia (Dr. Charles & Clara's second son)
1905–1906		~ Big rig built on Fairbank oil property in Oil Springs
1906	Jan. 20	~ **Henry Churchill Fairbank** is born in Petrolia (Dr. Charles & Clara's third son)
1908		~ Dr. Charles and Frank Carmen buy 158 acres in Elk Hills oil field, California

1910

1910		~ King Edward VII dies. Reign of King George V begins
1911		~ Dr. Charles runs as Liberal candidate in federal election but does not win
1912		~ J.H., 81, hands over business affairs to son, Dr. Charles
		~ Dr. Charles, Clara and their sons (John, Charles II and Henry) move into mansion with J.H.
1913		~ Barn at Fairbank Oil burns down and is rebuilt
1914 -1918		~ World War I
1914		~ Dr. Charles is reeve of Lambton County until 1918
	Feb. 1	~ **Jean Harwood** (future wife of Charles Sr.) is born in Moosejaw, Sask.

	Feb. 10	~ **John Henry Fairbank** dies in Petrolia at the age of 83
	Mar. 14	~ Fairbank gas gusher in Oil Springs is biggest in Canada
	May	~ Oil Springs Oil and Gas Company strikes even bigger gas well in Oil Springs
		~ Oil is discovered in Alberta's Turner Valley
	Aug. 4	~ World War I begins
1915		~ Imperial Oil builds a receiving station at Fairbank Oil
		~ Fairbank Oil is valued at $40,000
	Sept. 8	~ **Robert Theodore Fairbank** is born in Fairbank mansion, Petrolia (Charles & Clara's fourth son)
1916		~ Dr. Charles in the trenches at The Somme at age 58
		~ Dr. Charles and Frank Carmen lease their Elk Hills land in California to Standard Oil
1918		~ Electricity arrives at Fairbank Oil
		~ Fairbank family sells Stevenson Boiler Works to Canadian Oil for $7,000
		~ At Fairbank Oil, boiler room at Fairbank & Shannon rig is converted to a blacksmith shop
	Nov. 11	~ World War I ends
1919		~ Dr. Charles becomes mayor of Petrolia
		~ Oil is struck on Fairbank & Carmen land in Elk Hills, California
1920		
1925	Feb. 24	~ **Dr. Charles Oliver Fairbank** dies in Santa Barbara, California at age 66
		~ Charles Oliver II, 21, (Charles Sr.) now owns one-eighth of Fairbank Oil
1927	July 21	~ **Clara Fairbank** (Dr. Charles' widow) marries **Leo Ranney** in Fairbank mansion
	Nov. 1	~ **John Henry Fairbank II** dies in Petrolia at age of 26 (first son of Dr. Charles & Clara)
1929		~ Stock market crashes. The Great Depression begins
1930s		~ Fire destroys original rig at James property of Fairbank Oil in Oil Springs

		~ North James Rig, South James Rig and Orchard Rig are built at Fairbank Oil
		~ Clara Fairbank Ranney sells Bothwell oil properties
		~ Clara Fairbank Ranney sells main street portion of VanTuyl and Fairbank Hardware Store
1934–1938		~ Charles Sr. becomes reeve of Petrolia
1935		~ Fairbank family loses title to 158 acres of oil land Elk Hills, California to U.S. government
1936		~ King George V dies. Edward VIII becomes King
		~ Crown Savings and Loan merges with Industrial Mortgage and Trust Company
1937		~ King Edward VIII abdicates. George VI becomes king
1938–1942		~ Charles Sr., a Liberal, wins East-Lambton provincial byelection and is youngest Member of Parliament
1939	Sept. 3	~ World War II begins
1939–1945		~ World War II
1940		
1940	June 20	~ **Charles Fairbank II** (Charles Sr.) marries **Jean Harwood** in Moosejaw, Sask.
1941		~ Charles Sr. and Leo Ranney go to Australia to advise its government about expanding Australian oil production
	Sept. 1	~ **Charles Oliver Fairbank III** (Charlie) is born in Petrolia (first child of Charles II & Jean)
1943	Feb. 4	~ **Sylvia Jean Fairbank** is born in Petrolia (second child of Charles II & Jean)
1944		~ **Mary Edna (May) Fairbank Rock** dies in California at 75
1945	May 8	~ World War II ends
1950		
1950		~ **Leo Ranney** dies in California at age 66
1950's		~ Date unknown, the last of the horses is used at three-pole derricks at Fairbank Oil
1952		~ King George VI dies. Elizabeth II becomes queen

1955 Aug. 18 ~ **Patricia McGee** (future wife of Charlie) born in Ottawa
1956 July 28 ~ **Clara Fairbank Ranney** dies in Fairbank mansion in Petrolia at age 79

1960

1960 ~ Railway to Oil Springs is closed
~ Oil Museum of Canada opens in Oil Springs
1961 ~ Fire destroys the big rig at Fairbank Oil. East Rig and West Rig built
1962 Dec. 19 ~ **Henry Churchill Fairbank** dies in Petrolia at age 56 (third son of Dr. Charles & Clara)
1964 Jan. ~ Fairbank family loses $15 million law suit over Elk Hills oil land in California
1965 May 7 ~ Edward Phelps hands in his thesis, *John Henry Fairbank, of Petrolia* (1831 –1914) A *Canadian Entrepreneur*
1967 ~ Fairbank mansion is sold to Ron Burnie for $25,000
1969 ~ Charlie, (Charles III), works one year at Fairbank Oil
~ Charles Sr. now owns half of Fairbank Oil
~ Charlie urges father to try to obtain other half of Fairbank Oil

1970

1973 ~ Charlie returns to Oil Springs and is sole owner of oil property
~ Energy crisis ensues with oil embargo. Price of crude oil skyrockets
1974 ~ Charlie incorporates the name Charles Fairbank Oil Properties Ltd. and takes over the mortgage
~ Oil property has 70 productive wells, 70 inactive on 350 acres
~ Imperial Oil Receiving station on oil property closes
~ Charles Sr., Gary Ingram and others start planning an outdoor oil museum, (The Petrolia Discovery)

1980

1980 ~ Charlie moves into the 1888 farmhouse on Fairbank Oil and renovates for a year
~ The Petrolia Discovery opens
~ Charles Sr. is on the Discovery board, Charlie is on the technical committee

1981 ~ Mural sketched on the barn at oil property by Ann Evans of Wyoming during first Fairbank Lamb Roast

1982 Mar. 1 ~ **Charles Oliver Fairbank II** (Charles Sr.) dies in Petrolia at age 77

~ Charlie becomes owner of VanTuyl and Fairbank Hardware

~ Charlie joins the board at The Petrolia Discovery, stays 21 years

~ Ontario Petroleum Institute gives the Fairbank family the Award of Merit

1983 ~ Mural painted on chicken coop by Ariel Lyons for second Fairbank Lamb Roast

1990

1990's ~ Historic sites found on property: Tripp wells of 1850s and also the Black and Matheson "flowing well" of 1862

1991 ~ Two disposal wells using gravity-feed were added at cost of $250,000

~ Two adjacent oil properties bought from Don Matheson

1992 Feb. 4 ~ **Charles Oliver Fairbank IV** (Little Charlie) is born in London, Ont. (first son of Charlie & Patricia McGee)

~ Charlie receives Canada 125 Medal for his volunteering contribution to Petrolia Discovery

~ Two adjacent oil properties bought from David Baldwin

1995 ~ Oil property bought from Irv Byers estate

1996 Jan. 15 ~ **Jonathan Alexander Fairbank** (Alex) born in London, Ont. (second son of Charlie & Patricia)

~ Additional disposal well drilled at Fairbank Oil on west Baldwin section

1997 Aug. ~ J.H. Fairbank, James Miller Williams, Charles and Henry Tripp are inducted to the new Canadian Petroleum Hall of Fame in Leduc, Alberta. Charlie is the guest speaker to the audience of 800

~ Adjacent oil property bought from Paul Morningstar

~ Oil tank on Gum Bed Line is painted as Thomas the Tank Engine by Renée Ethier

~ Sue Whiting commissioned to paint some of the pump jacks with dinosaur faces

1998	Mar. 2	~ **Jean Fairbank** dies at the age of 84, in Petrolia
1999		~ Adjacent property bought from Phil Morningstar
		~ Dr. Emory Kemp and the industrial archaeologists from West Virginia Universtiy arrive at Fairbank Oil to study it in detail

2000

2000	Sept.	~ American Association of Petroleum Geologists tour Fairbank Oil
2001		~ Fairbank Oil's 140 the anniversary. Property is 600 acres with 350 wells
		~ Industrial archaeologists from West Virginia University conduct offical field school at Fairbank Oil
2002		~ Queen Elizabeth II Golden Jubilee
	Aug. 4	~ Fire badly damages Sunnyside in Petrolia, also known as the Fairbank mansion
2002-2003		~ Addition built on Fairbank farmhouse at Fairbank Oil Properties
2003		~ Charlie receives Queen's Jubilee Award for his volunteer work
		~ Charlie steps down from Petrolia Discovery board after serving more than 20 years
		~ A Sunnyside committee applies for heritage designation of mansion
		~ A Building Condition Assessment for Sunnyside is prepared by Goldsmith Borgal and Company ltd. Architects and BKL Engineering
		~ Barn with mural at Fairbank Oil is raised up and given new foundation, cement floors throughout.

PART FOUR
Asides

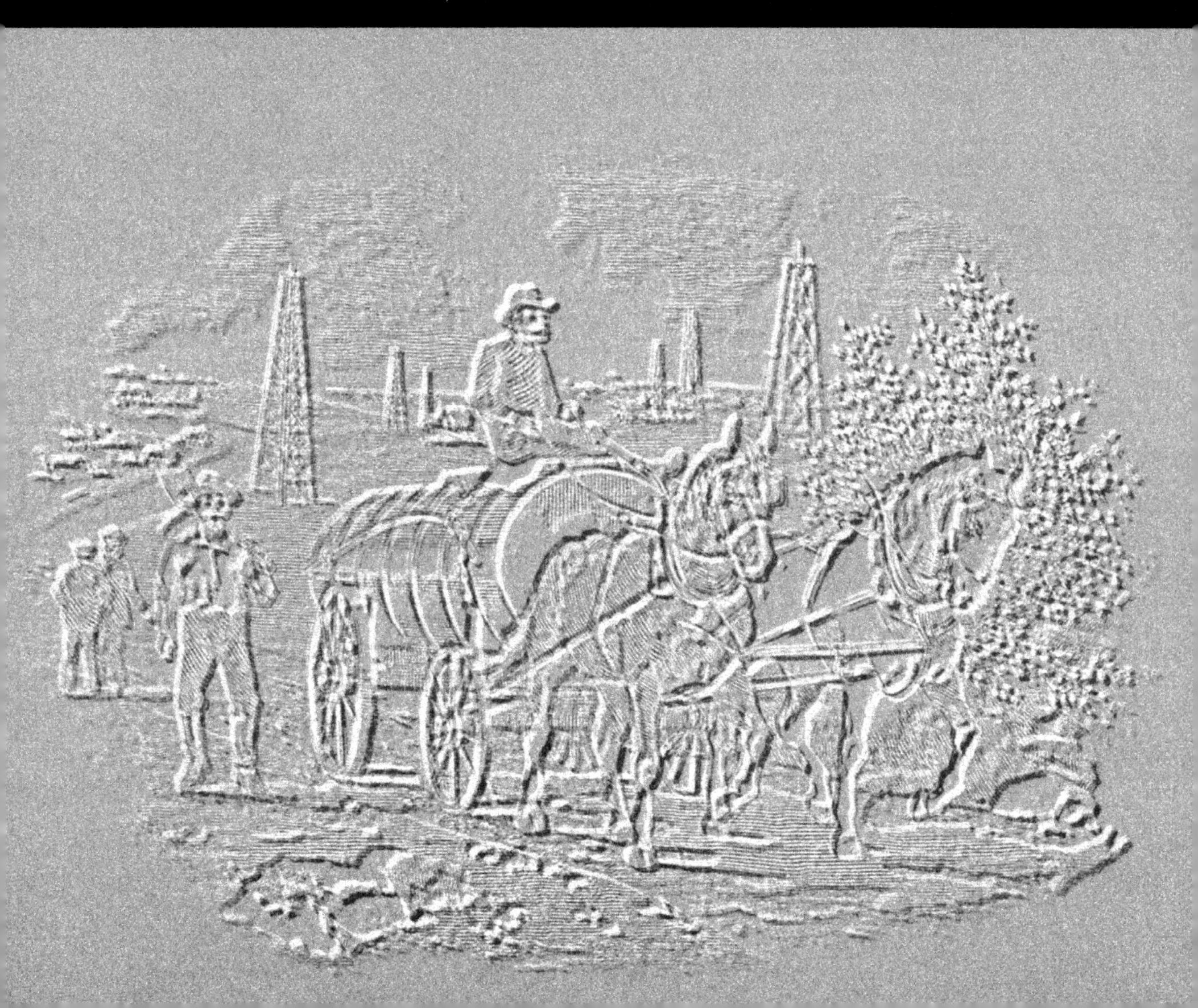

Chapter 10 – The House That Oil Built, The Fairbank Mansion

The Fairbank mansion has been a great curiosity to many. This curiosity goes right back to 1889, when two years of construction began. The owner and part designer was John Henry (J.H.) Fairbank, the largest single oil producer in Canada. Affectionately, he was known as "the father of the town" for his wide web of business interests that employed more than 400 people and a myriad of political roles that included everything from fire chief to federal Member of Parliament in Sir John A. Macdonald's government.

Part of the interest in the mansion comes from its five-acre site on the main street of Petrolia, above the treed slopes of Bear Creek's west bank. Standing three storeys high, the massive 22-room landmark at the eastern edge of Petrolia's downtown is impossible to miss.

The style of the building is arresting too. It's the only home in the town built in what's called the Queen Anne Revival design, a style characterized by its exterior of red sandstone, red brick and slate roof. Petrolia's St. Paul's United Church and the old Post Office, both on the main street, share this style as does Queen's Park, the provincial legislative building in Toronto.

Popular folklore has it that J.H. built the home, not for himself, but to please his wife, Edna. Despite his wealth, he was among the last of Petrolia oil barons to build a gracious home. The many attractive large homes of Crescent Park started going up in the 1870s. Jake Englehart, who owned the largest refinery in Canada, had purchased the Glenview mansion from Henry Rosenberg in the 1880s and gave it to his wife, Charlotte. As Charlotte requested in her will, Englehart donated this mansion to become the town's Charlotte Eleanor Englehart Hospital.

The Fairbank mansion was owned by

For 74 years, the mansion was owned by the Fairbank family

the Fairbank family for 74 years. J.H.'s grandson, Charles Oliver II (later known as Charles Sr.), was the owner in 1963 when a crowd of more than 3,000 people lined up to tour it on a Sunday afternoon. Another tour that year brought a crowd of more than 5,000. At the time of the first tour, it appeared the home would be demolished because Charles Sr. could not find a buyer. He had been trying to sell it since the death of his mother, Clara Fairbank Ranney, in 1956. In the intervening years, it stood fully furnished but unoccupied.

A second tour of the Fairbank mansion in 1963 brought a crowd of more than 5,000.

Long before he died in 1914, J.H. had told his family that if the mansion should ever become a burden they should rid themselves of it. This message was handed down the generations and by the time the home was owned by his grandson Charles, it had become quite a burden. The mansion harkened back to another era and was built in age of servants. Each room was equipped with bells for summoning staff; a panel of bells in the maids' sitting room indicated which room needed their services.

When the word spread in 1963 that the home was destined to be torn down, a small group of Sarnia citizens led by Mrs. Grant Smith were horrified enough to take some action. They arranged a meeting in Petrolia's Victoria Hall that gathered "key people of Lambton County, from the warden on down," reported Joan May in *The London Free Press* on April 27, 1963. "That day, Fairbank House was saved from the wrecker's hammer. Fairbank Adult Education Centre was born."

The centre was to become an educational retreat. Students would live at the house "to be under constant exposure of the topic under study, away from the distractions of every day life," according to the newspaper.

"...Charles Fairbank (Sr.) listened with interest to the citizen's plans – and made his offer. He would lease the house to them, rent free," she reported. "They could use the furniture which remained, rent free. They would pay the taxes and maintain the property. That was all.

"Meanwhile word had spread that the house was doomed. The Sarnia real estate agents who had been trying to arrange a sale were deluged with telephone calls – not from potential buyers, but from people who wanted to see the old house before it disappeared forever.

"They arranged for two guides and a

Sunday afternoon tour. They expected 500 people. Between 3,000 and 4,000 came…Last weekend a tour and tea were held to introduce Fairbank House, adult education centre to the community. The success of the previous open house was repeated – but even more so. Estimates placed the crowd at between 5,000 and 6,000 persons," wrote Ms. May, adding cheerily, "More than $100 was raised in donations."

Florence Moore was one of the thousands who took the tour. She was in Grade Six at Jubilee School at the time and distinctly remembers having a fit of giggles at the top of the grand staircase. A throng of people were climbing the stairs and she noticed that on a level below her, a big pink blob of bubblegum had dropped on top of a woman's dark brown beehive hairdo. "Some little kid must have been leaning over the bannister and dropped the gum," she said. The hilarious part for Florence and her friends, was that the woman never even noticed, which is understandable, for frothy hairdo seemed to be at least eight inches high.

These two afternoon tours were so wildly successful that the citizens group, known as Fairbank House, thought regular tours would raise money for the education centre's director. A brochure states the Residential Adult Education Centre was open "every day until September 20 from 1 to 5 p.m. Admission $1, Children, Free." Under the headline "See Historic Fairbank House" it included maps of how to reach Fairbank House from Sarnia and London.

The design of the mansion was "perhaps the last great creative act of J.H. Fairbank's long career."

The design of the Fairbank mansion was what historian Ted Phelps called in his thesis "perhaps the last great creative act of (J.H.) Fairbank's long career." He notes that J.H. "lavished many thousands of dollars and many months of careful and painstaking work to ensure that every part of the building was as it should be." The firm Peters, Jones & McBride of London, Architects helped prepare the plans and specifications. There are records of many subcontractors hired to work on the house. It is known, for example, that a Seaforth company called The Broadfoot and Box Furniture Company supplied 67 doors for $530. The ballroom floor, all family members knew, was made solid with an underlay of bricks.

The Fairbank House brochure was more effusive. "Surrounded by five acres of orchards, lawns and formal gardens, the home itself is built of red

"To get inside the Fairbank mansion was an event in those days. Some people made it their ambition."

clay bricks, all carefully packed in straw before shipment from Ohio. Timbers and hardwoods in the house were cut from the Fairbank farms in Brooke Township, Lambton County, and it was the owner's boast that all the wood was dried for use for a year prior to use. The woodwork was carved by hand and still retains its charming natural shade. The third floor of this stately mansion features a ballroom in which the cream of local society often gathered in days of bygone splendour. The entire structure has been carefully maintained and remains today as sturdy as when new."

Bertha Gleeson's memory of the house is from decades before there were any tours. She and her husband, Lew, had attended an event at Victoria Hall and as they were leaving, Lew noted that Clara Fairbank Ranney was leaving unaccompanied. Lew suggested they walk her home to the Fairbank mansion and Clara accepted his gracious offer. At her front door, she asked them in. Lew began to refuse, but Bertha swiftly elbowed him, whispering, "This is our chance!" and they entered together. "You have to realize," said Bertha. "To get inside the Fairbank mansion was an event in those days. Some people made it their ambition!"

Two years before J.H. died in 1914, his son, Dr. Charles Fairbank, moved into the mansion with his wife Clara and their three sons. Their fourth son, Robert, would be born in the mansion in 1915. Earlier, the doctor and his family lived in what is known as the VanTuyl House, or the yellow house, at 420 Warren Avenue in the Crescent Park neighbourhood. Dr. Charles lived only 11 years after his father's death. Clara continued to live in the Fairbank mansion with her second husband, Leo Ranney.

Fate would have it that Clara was the last Fairbank to live in the home. Her son, Charles Sr., and his wife Jean, did not particularly want to live in the mansion and raise their two children, Charlie and Sylvia, there. Properly maintaining a home of this size and its extensive gardens was simply too expensive. They knew they would pay dearly in both money and time.

"My father was a realist," says Charlie. "He knew who he was and didn't need to have a mansion to feel important. He was afraid that looking after the house would wear my mother out." For more than 50 years, their family home was at 425 Warren Avenue in the Crescent Park neighbourhood.

The mansion was officially named Sunnyside and others would refer to as Fairbank House but in the Fairbank family, it has always been called simply, The Big House. This name has something of a Texan ring to it, possibly an Americanism that the Fairbanks adopted.

The best description of the house can be found in a story Charles Whipp, wrote in *The London Free Press.* This is an excerpt from his longer story which appeared on November 28, 1959.

"The house is now owned by C.O. Fairbank, grandson of J.H., and who, while not a poor man, admits he cannot afford to live in it. Currently it is leased as a rest home. But the casual visitor could wander through its 20-odd rooms almost without noticing its four elderly tenants.

"Iron gates off Petrolia's main street admit the visitor's car up the long driveway to the main house. As the green and yellow cover of trees and vines part, the great mass of sandstone and brick assumes its shape. Slate roofs tower over 50 feet above the street. The meticulously laid brickwork curves in turrets and wings and the plate glass windows and storm sash curve faithfully with them. An ancient wisteria vine clambers over the broad front verandah knotting itself to the wrought iron railings.

"When the oaken door swings back, one is presented with a view of the centre hall, 60 feet long and 12 feet wide, with wainscoting of rubbed sycamore, beautiful as the day it was cut from the Fairbank farms in Brooke Township. To the left is the music room with its grand piano (a Steinway, of course) and beyond, the great staircase, with its hand carved bannisters, moves up to a vast landing where a picture window 12-foot square sends a flood of light over the gleaming sycamore. Five people could walk abreast up this staircase to the third floor ballroom where famous orchestras (Guy Lombardo's among them) once played to Petrolia's oil rich high society. Earl Grey, Lord Lascelles and (Indian poetess) Pauline Johnston are among the famous names entertained here.

"... the great staircase, with its hand carved bannisters, moves up to a vast landing where a picture window, 12-foot square, sends a flood of light over the gleaming sycamore. Five people could walk abreast up this staircase..."

"Fairbank has eight bedrooms, none less than 14 feet, and usually with adjoining dressing rooms and bathrooms. The walk-in closets have sliding doors of great weight, but which slide easily. There is none of the dimness and darkness in these rooms, characteristic of so many early houses. Some of the windows go from floor to ceiling and in the bedroom wings there are two and three windows in a row, providing excellent ventilation. Everywhere the woodwork of oak, ash, birch and maple, is like new. Cut from the woods of Lambton it was aged a year before the carpenters received it. The brickwork came from Ohio, each brick individually wrapped in wax paper.

"The attics of the third floor would, together, accommodate the average bungalow of today. They are filled with antique furnishings, occasionally protected by dust sheets, and with books, letters, documents, broken or discarded clocks; the accumulations of three generations of a great house and its families.

"In the basement one wanders through the wine cellar, the fruit cellar, the kitchen cellar, the laundry, the heating rooms, the coal cellar and the wood cellar. Two furnaces, one coal, the other gas, heat Fairbank house. A man can stand in their cold air vent. The basement is nine feet from its concrete floor to plastered ceiling. Construction throughout is 14-inch joists set on 12-inch centres. In the upper attic, heavy timbers support the roof in a barn-like construction."

By any measure, the mansion is enormous. Finding a proper use for it and more importantly, the finances to maintain it, was a perpetual problem. Wisteria House, as it was called as a home for the elderly, was not financially viable, nor was the education centre. There had been a few other uses too. In 1958, it served as a church. Fire had destroyed Christ Church Anglican on Oil Street and so the congregation moved to the Fairbank house for services until their church was rebuilt.

"I think the main reason I wanted to take the ballet was because it was in the Fairbank mansion and it was in the ballroom."

The Fairbanks opened the house in the summer of 1966 to have a bar there for Petrolia's Centennial celebrations. And it was likely 1966 when Pat Stephen, owner of Total Hair and Body Care in Petrolia, took ballet lessons in the ballroom.

"I think the main reason I wanted to take the ballet was because it was in the Fairbank mansion and it was in the ballroom," she said. It was a six-week summer program taught by Mrs. Tipperly, who always wore black tights despite the hot weather. Tap dancing was taught as well. The house was totally vacant of any furnishings and, despite the daylight, it was rather scary for Pat Stephen. "I remember going up those big stairs to the third floor," she said. "I always felt something was going to reach out and grab me."

The ballroom, was of course, perfect for dances. In 1964, two or three teen dances were held there and once, and sometime in the early 1950s, Clara opened her house to allow a nurse's association to hold a dance there.

The contents of the house also far exceeded any needs or desires of the family. In the fall of 1966, three separate auctions were held at the mansion: Sept. 29 for the house itself, Sept. 30 for the contents of the library and Oct. 1 for all furnishings. These last two auctions included items that had not belonged to the Fairbanks. Gardner Auctions Limited of London may not have been optimistic that the house could be sold. They ensured that readers of their newspaper advertisement were informed of any further auctions. They stated, "IMPORTANT NOTE – If the property remains unsold, certain structural detail will be sold for later removal – these include carved sycamore and other panelling, marble and tile fireplaces, cast iron railings, coloured glass panels, doors, lighting fixtures, bathroom fixtures and similar items not normally available."

The house did not sell at the auction but Charles Sr. held on to it. The fixtures and panelling remained untouched. Finally, the house was sold in the summer of 1967 to Ron Burnie, an instrument mechanic with Polymer in Sarnia. He and his wife, Jessie, and their five children then moved in. During the early 1990s, the Burnies had an antique store that occupied the large front foyer and the library. Four apartments were added to the house with one built over the carriage house to the north of the mansion. A fire there destroyed the roof. Electrical problems in the mansion led to the closing of the apartments.

The Town of Petrolia has attempted to buy the home from Burnie. It first took out an option to buy the house in 1983. No plans were revealed about the intended use of the house. An *Advertiser Topic* story dated August 17, 1983 said "Ironically enough, the Fairbank family offered the town the building and the property for $25,000 about 15 years ago but the town council then said it couldn't afford the expense."

During the 1980s, the town paid for a real estate assessment and the house was valued at $425,000 at the time. The town offered $500,000 and the offer was refused. A good portion of this money was to come from the federal govern-

The contents of the mansion also far exceeded any needs or desires of the family.

ment for improvements. Negotiations began anew in the early 1990s, again with the idea of federal help. However, federal assistance could not be found. This offer was also refused.

Tragedy struck on August, 4, 2002 when fire swept through the third floor, badly damaging the billiard room, den, ballroom, two attics and the roof. Ron Burnie and his wife, Jessie, had escaped unharmed. Smoke from the flames and water from the firefighters' hoses have scarred the main floor and second storey. Although the mansion's damage is extensive, the house has not been lost forever. A new roof was constructed to preserve what's there.

Charlie, Sylvia and their parents would spend their Sunday evenings at the Big House.

In the summer of 2003, the future of the mansion was still in question. To find out how feasible it is to restore the house, the town of Petrolia has turned to experts who rebuilt Victoria Hall after its fire in 1989 – Architect Philip Goldsmith and his Toronto firm of Goldsmith Borgal & Company Ltd. and Clive Barry of BKL Engineering in Sarnia.

Petrolia has formed a Sunnyside committee, headed by Mary Pat Gleeson, and it has applied to the Ontario government to obtain a heritage designation for the house. "We believe we are stewards, caretakers and guardians of our past," she said. "It is for those who follow that we are working to create awareness and to preserve tangible symbols of one of the most important events in Canadian and world industrial history – the discovery and exploitation of crude oil."

When their grandmother died, Charlie Fairbank was 15, and his sister, Sylvia, was 13 years old. They remember that she had a stroke and was bedridden shortly before her death. Their mother, Jean, nursed her and private nurses were hired as well. Margaret Robson, whom the family had known for decades, did light housekeeping and was a companion to Clara Fairbank Ranney. Earlier, she had been the nanny to their father, Charles Sr., and possibly she was nanny to his brothers, John and Henry too.

In the early 1950s Charlie, Sylvia and their parents would spend their Sunday evenings at The Big House. "My grandmother had a television before we did," says Sylvia. "We always went there to watch the Arthur Godfrey Variety Show." The family tradition was to also spend New Year's Eve there.

The house may have been amazing to many, but to young Charlie and Sylvia it was familiar and thereby not so wondrous. Sylvia does recall that her grandmother had exquisite taste and what she remembers more than anything were the spectacular gardens on the five-acre property. Throughout the mansion, vases of freshly cut bouquets were everywhere. "My grandmother arranged the flowers herself and was very skilled at it," she recalls. "They were beautiful." Clara took immense pride in her garden and apple orchard but it was her son, Charles Sr., who did all the work in maintaining them.

In Sylvia's memory, her grandmother was dressed in black from head-to-toe the whole time Sylvia knew her and always she was seated in the library. For Sylvia, visiting was considered a duty, for the grandmother seemed cool and somewhat distant. Each year, though, her grandmother brought her back a new doll from California. After one trip, she brought back a cowgirl outfit for Sylvia and she instructed her husband, Leo, to put a nickel in each pocket.

Only once did Sylvia spend the night in The Big House, she was in Grade One. Her mother was briefly in hospital, her brother was away, and so Sylvia, who was nursing a broken arm, went with her father. "I didn't sleep a wink," she says. "I could hear squirrels in the attic and I was terrified. I cried so much my dad took me to sleep at our friend Dr. MacCallum's home the next night."

As a small boy of about 10 years old, Charlie and his friends Fog Wilson and Ron Atkey, visited the gardens of the mansion regularly. Together the boys raised chickens there and sold the eggs. They also made apple cider with apples from the estate's orchard and sold it at the Petrolia Fall Fair.

Near the garden, Charlie's grandmother kept a decorative wooden wheelbarrow that was painted a shiny red. It had a large steel wheel at the front. One day Charlie made the mistake of using it to hold chicken droppings when he was cleaning out the hen house. "Then I received a summons to go see my grandmother who was not amused that something so decorative could be used so vilely. It was, I believe, the last time that wheelbarrow was used to actually carry anything."

Charlie had a regular chore at his grandmother's house. It was to keep the fireplaces supplied by refilling the wood box. This meant frequent trips to the basement to chop the wood his father had loaded there. For a small boy, the vast basement was a scary place. He would charge down the stairs

For a small boy, the vast basement of the mansion was a scary place.

After staying in the mansion, one memory stood out for him - the polar bear rug, complete with head, teeth and claws.

as quickly as he could, then flick on the light. To drown out the creepy sound of termites chewing, Charlie would chop the wood making as much noise as he could. "That way," he says, "I could keep all the other noises at bay."

Like Sylvia, he too slept only one night in the mansion. He slept in Leo Ranney's room, with a very high bed and the memory that stood out for him was the polar bear rug on the floor, complete with head, teeth and claws. Finding out that bear fur is coarse, not soft, was a big revelation. And to show how fickle memory can be, Charlie swears it was The Ed Sullivan Show the family watched together at his grandmother's, not The Arthur Godfrey Variety Show.

In the 1930s, there was a break-in at the mansion. Clara and Leo were wintering in California, as they often did, and Charles Sr. routinely dropped in at the mansion to check on things. At the time, he was living across the street in the apartment above The Little Red Bank. When he entered the living room, he noticed the rug was not in its proper place. As he moved it, he saw that someone had created a big hole in the floor. This hole led directly to the wine vault below, a room that had a steel door and a combination lock. On further inspection he noted that a good quantity of liquor had been stolen.

He also noticed that a substantial amount of liquor remained and he reasoned that the thieves would be back to take more. Charles Sr. rounded up some friends to camp out in the wine cellar and catch the returning thieves. As it happened, these friends decided to open a few bottles for themselves. A boisterous party ensued and they made so much noise that thieves or anyone else could have heard them at some distance. The identity of the thieves has forever remained a mystery.

The Fairbank Mansion

This is the house that oil built. Officially, it was named Sunnyside but in the Fairbank family it was simply called The Big House.

Fairbank Family photo

The Gardens of the Mansion

With five acres of land surrounding the mansion, there was plenty of room for the extensive flower garden which flourished under Clara Fairbank's direction. The wisteria at the front of the mansion was so luxurious that the home was named Wisteria House when it was converted to a nursing home.

Fairbank Family photo

Fairbank Garden

Clara Fairbank and unidentified child stand at the garden gate. Fresh bouquets were everywhere in the house.

Fairbank Family photo

Reflecting craftsmanship of another era ...

Surrounded by five acres of orchards, lawns, and formal gardens, the home itself is built of red clay bricks, all carefully packed in straw before shipment from Ohio. Timbers and hardwoods in the house were cut from the Fairbank farms in Brooke Township, Lambton County, and it was the owner's boast that all the wood was dried for a year prior to use. The woodwork was carved by hand and still retains its charming natural shade. The third floor of this stately mansion features a ballroom in which the cream of local society often gathered in days of bygone splendour. The entire structure has been carefully maintained and remains today as sturdy as when new.

Fairbank House is now considered far too large for family occupancy, and has been leased by a local group of citizens as a non-profit residential adult education centre. The house has been found adequate for institutional use without alteration, so that its attraction to visitors is unchanged. The citizens group, also known as Fairbank House, is currently seeking the financial base necessary to support an adult education director and a program of live-in conferences, seminars, workshops, etc. Pending organization of this program, district community groups are invited to hold activities in the House. Directors of the House will be present to provide guided tours and commentary for visitors.

How to get to FAIRBANK HOUSE

Fairbank House is located on the North side of the main street, Petrolia Street, approximately 1 mile west of Highway 21. It is at the top of the hill west of the bridge over Bear Creek. Approached from the west, the house is at the east end of the Business section of the town.

TO REACH PETROLIA:

From Sarnia: 15 miles east along Highway No. 7, to Reece's Corners; then turn right (south) onto Highway No. 21. 7 miles to Petrolia. Turn right (west) onto Petrolia St.; 1 mile to Fairbank House.

From London: 47 miles west along Hwy. 22 and 7 to Reece's Corners; then turn left (south) onto Hwy. 21; 7 miles to Petrolia; turn right (west) onto Petrolia St.; 1 mile to Fairbank House.

Petrolia is 6 miles north of Oil Springs, site of the Oil Museum of Canada.

See Historic

FAIRBANK HOUSE

Residential Adult Education Centre
PETROLIA, ONTARIO

OPEN EVERY DAY UNTIL SEPT. 20th
1.00 to 5.00 P.M. D.S.T.
ADMISSION: Adults, $1.00. Children, Free.

Built in 1890 by John Henry Fairbank, oil magnate, banker, and politician, the Fairbank home symbolizes the solid prosperity of the town of Petrolia during three decades as the centre of the oil industry in Canada.

This house is an outstanding example of the craftsmanship and architecture of an era of low building costs. It was the largest private home within the County of Lambton and no expense was spared to make it one of the finest in Ontario.

John Henry Fairbank & Petrolia

John Henry Fairbank, builder of Fairbank House, rose from poverty to wealth in the oil industry a century ago. Born at Rouse's Point, New York State, he emigrated to Canada as a penniless young man of twenty-two, settling at Niagara Falls. There he married in 1855 Edna Crysler, a member of a well-known United Empire Loyalist family. He farmed, surveyed, and sold insurance, until he joined the crowd of oil prospectors after a surveying trip to Oil Springs early in 1861.

Oil Springs was then entering a five-year boom following notable oil discoveries. Fairbank bought a half acre of land from James Miller Williams, who in 1858 had drilled the first commercially producing oil well in North America close by. Fairbank put down a well of his own, struck oil, and after sinking more wells, built a tiny refinery, like many then in operation.

Fairbank prospered from his oil operations, and in 1865 he moved seven miles north to Petrolia, to develop newer oil properties which became the basis of his fortune. He was one of the leaders of the crude oil producers of Petrolia during their economic contest with the refining interests located at London, Ontario, at a time when each group sought to dominate the industry.

In 1865 Fairbank founded the hardware business later known as VanTuyl and Fairbank, and together with L. B. Vaughn, founded in 1869 Petrolia's famous "Little Red Bank," which survived until 1924 as a private banking institution. The building still stands; it is the oldest commercial structure in the town. Here the oil magnates gathered daily from 1884 to 1898 to trade in stocks of oil on the Petrolia Oil Exchange. Later in life Fairbank also owned a boiler factory, and manufactured fire extinguishers.

Known in his own lifetime as the "Father of Petrolia," Fairbank worked hard for the welfare of the town. He served as Reeve between 1868 and 1870, built the town's first jail, and was Fire Warden from 1874 to 1889. Between 1882 and 1887 he represented his constituency of East Lambton as an opposition Liberal member of the Canadian House of Commons, at Ottawa. He was a generous supporter of Christ Anglican Church, Petrolia, and many other worthwhile community groups and enterprises.

Mrs. Edna Fairbank died in 1896. After John Henry Fairbank's death in 1914, many of his enterprises were carried on by his son, Dr. Charles Oliver Fairbank, himself a noted oil producer, sportsman, and army major. The family is now in its fourth generation in Petrolia.

PETROLIA, ONTARIO, was established as an important centre of oil production in 1865, and was incorporated as a village in 1866. It become a town in 1874, when its population was around 5,000. For about twenty-five years it was wholly dependent upon the oil industry for its existence. As the boom town of the 1860's matured into the prosperous centre of the 1870's and 1880's, other industry was attracted, which largely sustained the town when the oil production fell off early in the 20th century. Imperial Oil Ltd. (Esso) was founded in Petrolia. Today its wells, among the oldest on the continent, still pump oil from the early fields. Petrolia is also the hub of an excellent and flourishing farming district.

.... near North America's first oil well and first petroleum industry

Touring the Mansion

In 1963, a crowd of 3,000 lined up to tour it on a Sunday afternoon. Another tour that year brought a crowd of more than 5,000. This is the tour guide published at the time.

Fairbank Family papers

Chapter 11 – VanTuyl and Fairbank Hardware Store

Before Canada became Canada, before Petrolia became an incorporated town, Fairbank Hardware was already doing a brisk business on a muddy street of this shanty town of oil derricks.

It has earned a distinct place in Lambton County's history - no other family retail business is rooted in the 19th century and continues to greet customers in the 21st century. Opening its doors in 1865, this pre-Confederation family concern is believed to be the oldest independent hardware store in Ontario and a rarity in Canada.

The store, at 394 Station Street, is just a half block north of Petrolia's main street and kitty-corner to the town library housed in the former railway station. Surviving two world wars, the great depression, two fires, occasional thieves, numerous recessions, floods and a tidal wave of change in technology is no easy feat. The business has been passed from father to son three times since John Henry Fairbank founded it and these sons have all been named Charles Fairbank. The current owner is better known as Charlie, and he took over his great-grandfather's store in 1982. There's no mysterious secret to the store's longevity he says. "It just means that, for a mere 140 years or so, we've had to continually adapt and change."

Anyone seeing VanTuyl and Fairbank Hardware today is more likely to be struck by what hasn't changed. On the outside, there are walls of board and batten and a profusion of wild hollyhocks that have been at the store's alley entrance as long as anyone can remember. Together, they give the air of a bygone era.

If the store's exterior doesn't look like a shiny new Canadian Tire, the impression is even more pronounced inside.

If the store's exterior doesn't look like a shiny new Canadian Tire, the impression is even more pronounced inside. In the main office, customers pay their bills next to 19th cen-

"We're floating away! We're floating away!," he yelled into the phone.

tury photograph of an odd looking man standing with two cannons. Beneath the photo someone has written "VanTuyl and Fairbank Debt Collection Department". Above, there is a sign handpainted by Fred Bicknell. "Use our easy payment plan," it says, "100 per cent down, nothing more to pay." On the far wall is a framed newspaper from 1890 advertising the mortgage sale of a Petrolia refinery. And another wall was long adorned with a collection of photos that says, "Presented to John H. Fairbank by his employees, Christmas 1895." Two fly swatters hang on a nail and there's a container of grocery bags, thoughtfully donated by customers, for taking home the purchases from the hardware store.

The plywood floor can't hide the fact it has endured a lot of traffic in work boots. The computer-topped desks are surrounded by old wooden cabinets and shelves stuffed with well-thumbed catalogues of parts dating back to the 1920s. The two doors leading from the main office are clearly from another age – a scant two feet across and only six feet high. One of the employees, Tim Boot, measuring in at six-foot six, suffered numerous bruises before adjusting himself to the small door frame. This building has been part of Van Tuyl and Fairbank Hardware since the early 1870s.

The store is actually an unheated warehouse, or more precisely, a collection of large sheds brought together with roofing. Huge rough timbers support the roof, some are charred from a long forgotten fire, and some are still covered in bark with a display of extension cords hanging on nails. Various floorings appear in different sections of the store, sometimes unevenly - cement, plywood, wooden plank and, in the steel room, it's dirt.

Occasionally, the building has flooded. "Only when it rained," said Charlie. Over the decades, the town has literally built up around the store, so the rainwater flowed down into it. A short but particularly violent summer storm in the early 1990s caused a deluge in the store. Sylvia Fairbank, Charlie's sister, received an emergency call for help from store manager Ron Brand. "We're floating away! We're floating away!" he yelled into the phone. "Everything was under about a foot of water," she remembered. "We were wading in our Wellies trying to lift boxes up to drier shelves." Not long afterwards, the town minimized the elevation problem by providing better drainage.

The product line has not varied much in the last few decades because the four main groups of customers haven't changed. Stock is chosen to address the problems of industries, farmers, general consumers and those who operate the 700 heritage oil wells in the Petrolia area.

The vast majority of customers are decidedly male. Women may have been granted the vote and even pay equity during the store's time but in this store of heavy hardware it remains a man's world. It does not stock housewares, cookwear or any gardening gear that could be construed as pretty. Such items are only to be found in Petrolia's other hardware store, a place the staff occasionally refer to as "the girlie hardware store" when no one else is in earshot.

While some businesses devise complex marketing plans each year, Van Tuyl and Fairbank Hardware's advertising is next to nil. Its main advertising is in the programs for the Petrolia Community Theatre productions. Anyone attending one of the six sold-out performances of The Sound of Music in April, 2003 would likely have seen the hardware store that read: "Since 1865, THE Place to Shop for A Few of Your Favourite Things. We thread pipe. We cut steel. And we offer a stunning selection of gum boots and sump pumps."

The Saga of the Store

When John Henry Fairbank first built his store in 1865 it was just west of Bear Creek and it originally carried groceries and liquor. Within a decade, it carried a great number of high quality items for the Victorian home ("girlie items") but it specialized in hardware and oil well supplies. In 1869, the store was known as Fairbank & Bennett and Alexander B. Bennett was his partner. On July 1, 1874, he entered a contract with a new partner, Major Benjamin VanTuyl, who still carried his military title from the American Civil War. The store was renamed once again, this time to VanTuyl and Fairbank Hardware.

"We thread pipe. We cut steel. And we offer a stunning selection of gum boots and sump pumps."

Producing oil was Fairbank's primary business and it was to become his magnificent obsession. He arrived in Canada as a near penniless American surveyor who had taken a Canadian bride in Niagara Falls. Like thousands in his day, he was lured to Oil Springs by wild tales of striking it rich in the oil fields. He established Fairbank Oil Properties in Oil Springs in 1861 and by 1865, had amassed enough money to invest in the store. His oil business was really in its infancy but by 1890, he would be known as the largest oil producer in Canada producing 25,000 barrels a year.

Sometime in the early 1870s, Fairbank moved his store to the corner of the main street and Station Street. At that time the store was massive, and stretched all the way back to Railroad Street making it the largest hardware store west of Toronto. VanTuyl and Fairbank reduced their retail competition in 1877 by purchasing a store in the east end of Petrolia owned by Thomas Henry Smallman. Three years later, Smallman became one of the original 19 oil refiners who created Imperial Oil in London and is credited for starting the first chemical company.

At that time, the store was massive - the largest hardware store west of Toronto.

Reaching their customers in the age of horse and wagon was an incredible challenge. The road between Petrolia and Oil Springs was so deep in thick mud that it was known to cripple horses. Fairbank concluded that if oil producers could not get to his store he would take his store to them. Using the railway to ship the goods he opened satellite stores in Oil Springs and Bothwell. Dates on when this began and when it ceased are unclear.

The story of the store is interwoven with the epic events of the area and the world. Before the turn of the century, the economic foundations of the whole town were beginning to collapse. Booming oil fields were no longer gushing, they slowed to a trickle. Imperial Oil abandoned Petrolia taking its headquarters to Sarnia and the Petrolia Oil Exchange had gone out of business.

Adding to these business woes was the death of VanTuyl in 1900. J.H. Fairbank was 70 by then and VanTuyl had been his friend and valued partner for 26 years. As a measure of respect and gratitude, his name has never been removed from the store.

In the early decades of the 20th century, the fortune of the Fairbank family diminished considerably. J.H. Fairbank had backed a loan for the Petrolia Wagon Works, a large industry which was always financially wobbly and buckled under the final deathblow - the arrival of the automobile. In 1920, the family paid out $210,900 to the bank because of the outstanding loan. The famous stock market crash of 1929 and the ensuing depression further diminished the family fortunes. Also, the oil revenue from the area had shrunk. The sum of these events had implications for the hardware store. Feeling the need for more money in 1932, the family sold the principal Fairbank and VanTuyl Hardware store fronting on the main street.

Streamlined for Leaner Times

The remaining portion of the business was contained at the back of the building in the freight sheds. It was here that boxcars of goods had been unloaded from the railway branch line and it is here, just north of the alley, that the store remains.

The store's past helps explain its present and in today's world of superstores and retail chains, VanTuyl and Fairbank is as unique as a thumbprint. Each generation of Fairbanks has left its own stamp on the store. Charlie Fairbank Sr., (1904 -1982), had never intended to be involved in the hardware business. Instead, he had studied petroleum engineering at the University of California in the 1920s. He was on his way to a career in South America when his mother asked that he return to Petrolia for a year to straighten out the family businesses. He took over the store and oil field and stayed for the rest of his life.

In the 1930s and 1940s, he shifted the store's focus to water well casing and sold it across the province, later expanding the market to the Maritimes. By the 1960s, other larger suppliers muscled in on the market and VanTuyl and Fairbank moved back to general hardware.

Unlike his father, (Dr. Charles Fairbank) and even his grandfather, (J.H. Fairbank), Charlie Fairbank Sr. managed the store on a day-to-day basis, departing only when he served as the youngest Liberal member of the provincial parliament from 1938 to 1942. Two loyal employees became almost synonymous with the store. Elizabeth Hoban did the bookkeeping and Bill Hackett kept track of every nut and bolt in the warehouse. Collectively they worked at the store for 90 years.

One of the best remembered stories from the early 1960s is the tale of the Fairbank electric teapot. These round ceramic teapots contained a heating element that would boil water within minutes. It was a sideline business for Fairbank. Petrolia church groups were paid to assemble them and they were carried and distributed through the hardware store.

They proved to be just the thing needed for countless Canadian kitchens and they were so popular that they

The Fairbank electric teapot was just the thing needed for countless Canadian kitchens.

were to be listed in the Eaton's Christmas catalogue. Everyone was elated.

Soon afterwards a man from the Canadian Standards Association walked into the VanTuyl and Fairbank Hardware to test them. He decided the teapot should be able to boil water effectively for an entire hour. This was impossible. It boiled dry, so he declared the wattage had to be reduced. Once this was done, the teapot could no longer live up to its claim of boiling water within minutes. Without the quick boiling time, there was no interest in the teapot.

Financial records were handwritten in enormous books which looked liked they belonged in a story by Charles Dickens.

There's an epilogue to this tale. Years passed and Fairbank met a retired fellow from the Canadian Standards Association who clearly loved his original teapot. He made it his personal crusade to win the necessary approval for a short boiling time. He did not succeed.

To this day, boxes and boxes of teapots without heating elements sit piled beyond the steel room of the store. The original teapot had been sold with a guarantee. If the ceramic broke, the owner just had to send $2 and the heating element back to the store and it would be installed in a new teapot and sent back. Many did. And even now someone will occasionally wander into the store and inquire about them still.

A New Generation Ushers in the Millennium

Well into his seventies, Charlie Fairbank Sr. was still striding off to the store each day. Upon his death in 1982, his son Charlie took the helm. Like his great-grandfather, it's the business of producing oil that has really captured his heart. For well over a decade, he relied heavily on the capable management of Ron Brand who expanded the market with new lines and new customers.

It was also Brand's unenviable job to oversee the introduction of new technology. The store's financial records had always been handwritten in enormous books which looked like they might have come out of a story by Charles Dickens. Like most businesses, VanTuyl and Fairbank did not find the transition to computers an easy one.

In 1998, the reins of management were passed to David Taylor, who had been on staff for nine years. And in November, 2001, Theresa MacDonald made history by becoming the first female to manage Van Tuyl and Fairbank Hardware in all its 136 years of operation. As a sign of the times, customers rarely ask her if they could "speak to the manager". The men working at the store find it highly entertaining that sometimes a customer will phone and say he prefers to talk to a man about his hardware and they have suggested that Theresa try deepening her voice. She supervises three full time warehouse staff, a part-time employee in the office, and two students.

Living in an age when people can shop the malls and superstores every day of the week and almost any hour of the night, the hours kept by VanTuyl and Fairbank Hardware can be a bit of a shock to the uninitiated. It closes for lunch each day from 12 noon until one o'clock. Charlie makes no apologies and says the customers benefit from a happy, well-nourished staff. He points out that the store is not immune to change, in fact, until the 1960s, the store used to close each Wednesday afternoon and be open all day Saturday. Today, the store locks up at noon on Saturdays and is ready to greet customers at 8 a.m. Monday. It closes at 5 p.m.

Of course, there have been exceptions and Charlie remembers his father often abandoning his Sunday dinner to rescue some poor soul in dire need of a sump pump or pipe fitting. More recently, Victoria Hall has made some after-hours emergency calls to VanTuyl and Fairbank to replace lighting fixtures before the curtain opened on a performance.

Over the years, in the eras of both Charlie and his father, there have been a number of late night calls from the police reporting break-ins at the store. A big black safe was added to the office about 1870 and it bears the scars of various attempts to open it by those who were not privy to know the combination. One enterprising thief actually took a cutting torch from the shelves of products to vainly try to open the safe. Another thief helped himself to the broad axe, usually used for cutting rope, in hopes of whacking open the safe. He succeeded in only knocking the dial off. To date, no one has succeeded in breaking open the safe and if they did, they would find a sheaf of paper but little cash.

A burglar alarm has certainly helped in recent decades. Some kids once thought they were quite clever, choosing to break in during a power outage, only to find that the burglar alarm had a battery back up. An unofficial neighbourhood watch has

Charlie makes no apologies for closing the store at lunch hour and says customers benefit from a happy, well-nourished staff.

helped too. Pat and George Stephen live on the main street, above Pat's business, Total Hair and Body Care. Their second storey balcony overlooks the alley where VanTuyl and Fairbank has its entrance.

It is now not uncommon for the well-dressed patrons of Victoria Hall's summer theatre to saunter in.

On Halloween in 2001, the couple Pat and George Stephen happened to be on their balcony, relaxing in their hot tub around midnight. That's when they heard noises. Peering through the lattice of their privacy screen, they could make out the shape of two shadowy figures at an unused door of the hardware store. The Stephens quickly realized these shadowy figures were trying to jimmy the door and break in. Though she was in the buff and in the hot tub, Pat had the presence of mind to yell out, "Hey!" and the thieves bolted like scared rabbits. Some felons are more discreet. Years ago, during the period when Ron Brand was manager, a left-handed pillar of the community had been known to switch pairs of gloves in the store so that he would be purchasing two left gloves.

As time marches on and the days of a family hardware store drift further into the past, VanTuyl and Fairbank becomes even more of a curiosity. It's now not uncommon for the well-dressed patrons of Victoria Hall's summer theatre to wander in along with those who need some steel cut, pipe threaded or a pound of nails.

To the amusement of Charlie and the staff, telemarketers sometimes call the store wishing to speak to Mr. VanTuyl.

The Biggest Hardware West of Toronto

VanTuyl and Fairbank Hardware once occupied these buildings on Petrolia's main street. The horse and wagon at the left of the photo are on Station Street. The store was drastically reduced to its present size in the 1930s when Clara Fairbank sold these main street buildings J.H. is seen here in a light suit with his son, Charles, to his right. The photo was likely taken about 1910.

Fairbank Family photo

VanTuyl and Fairbank Satellite Stores

Along with the main Petrolia store, VanTuyl and Fairbank Hardware had satellite stores in Oil Springs and Bothwell. It's believed this photograph was taken inside the Oil Springs store.

Fairbank Family photo

VanTuyl and Fairbank Hardware in Oil Springs

This shot shows the exterior of the store in downtown Oil Springs. Date unknown.

Fairbank Family photo

VanTuyl & Fairbank
HARDWARE
Oil & Salt Well Supplies, Fittings, &c &c
PAID
$ 47 50

An Early Logo

The dates of the two satellite stores are unknown but the store's logo on this bill mentions stores in Oil Springs and Bothwell and the bill is dated June 7, 1924. The drawing from the store's original logo has been adapted for the barn mural at Fairbank Oil.

Fairbank Family papers

Petrolia 1874

In 1874, Petrolia buildings still had not embraced the permanence of brick. This shot from the main street overlooks VanTuyl and Fairbank Hardware.

Fairbank Family photo

In the Artist's Eye

Ann Evans, formerly of Wyoming, painted this picture of the hardware store when the hollyhocks were in bloom.

Fairbank Family papers

Chapter 12 – How Oil Found Its Way to Oil Springs and Petrolia

By Claudia Cochrane,
Cairnlins Resources Ltd.

When people refer to the story of oil, the beginning is usually set in Oil Springs, Ontario in the 1850s. A geologist's version of the tale opens about 375 million years earlier and the geographical setting is a very big chunk of the globe.

Geology tells us that beneath the earth's surface lay a number of rock formations layered on top of each other. Each formation has a name, a depth and certain characteristics. The oil in Oil Springs is found in the limestone formation named the Dundee. These limestone rocks were deposited in a time period known as the Middle Devonian, approximately 375 million years ago.

These particular rocks were deposited on the eastern margin of a large continent that straddled the equator. This ancient continent, sometimes called The Old Red Continent or Laurasia, was composed of land that is now North America and Greenland as well as parts of Russia, Northern Europe and the British Isles.

For a geologist, the story of oil begins about 375 million years ago.

During the Middle Devonian, a group of sea animals lived on the margins of a clear, brightly lit, very shallow sea that lapped over the shores of the continent. These shallow seas, or basins, of long ago are unknown to us in the modern world. The sea animals, or "critters" as some palaeontologists fondly call them, included corals, brachiopods, clams and sea lilies. The depth and breadth of the sea often changed and occasionally it even reached far inland to the very centre of the landmass.

When the sea animals died, their soft parts were eaten by scavengers or they rotted with oxidation. Their hard inedible shells, however, were scattered on the sea floor. Every now and then, the

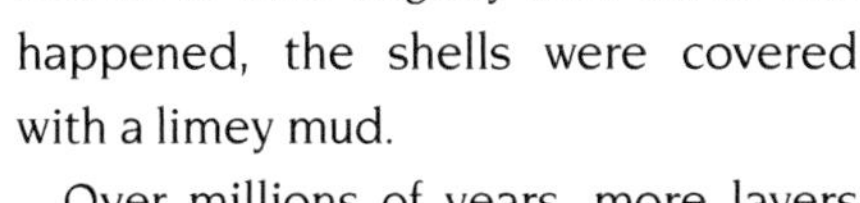

sea level rose slightly and when this happened, the shells were covered with a limey mud.

Trapped by non-porous limestone, the oil would be released from its rocky tomb by enterprising human beings.

Over millions of years, more layers of mud and sediment were deposited on top of the shells. The shells and sediment were slowly transformed into rock by the action of heat, time and pressure over time. This rock, known as limestone, was not totally solid. The odd shape of the shells created numerous small gaps and holes. Many of these gaps became interconnected and the result was a rock that looked like a sponge. The indentations, or pores, in this reservoir rock would later be filled with oil.

While this was happening in the shallow area of the basin, another story was unfolding in its deeper part. A large number of microscopic plants and animals floated in the warm brightly lit upper layers of the deep water. When these organisms died, their tiny bodies rained down to the basin floor. There, they collected and were preserved, soft and hard parts alike. During Middle Devonian, this part of the sea was a stagnant, low-oxygen environment, which was uninviting to most other critters, even scavengers. These dead organisms mingled with incoming fine-grained sediments and were covered. Over the millennia, these deeper layers accumulated and, like their shallow water neighbours, were eventually turned into rock by time, pressure and heat. The organic matter from the plants and animals was distilled to produce natural gas and oil within the fine-grained rock we call shale.

Both the porous limestone reservoir and the oil-filled source-rock shales remained separate and undisturbed over the eons. Then the earth began to move, causing the two groups of rocks to fracture and tilt. This created pressure and forced the oil to squeeze out of the shale. As it emerged from the rock, it migrated upwards, following the angle of the cracks and through the overlying porous rock. It kept migrating until it was stopped by solid non-porous rock. It was here that it became trapped in the limestone, waiting to be released from its rocky tomb by enterprising human beings.

And there's another part of the story. About 10,000 years before humans were investigating oil in Lambton County, glaciers swept down from the north. The glaciers acted like bulldozers, scraping away much of the rock and later depositing 30 feet of loose clay, sand and debris. This debris, sand and loose clay formed what is known in Lambton County as Brookston Clay. The oil below the Brookston Clay continued to seep upwards, using fractures for travel routes. Much of the oil collected in the sand of the Brookston Clay but some journeyed all the way up to the soil at the surface.

Once it hit the surface, the natural gas and lighter oils evaporated. This left a thick, smelly, tar-like substance, called gum, around Black Creek in the Enniskillen swamp. And those enterprising human beings bestowed the smelly gum with new name – they called it Black Gold.

This left a thick, smelly, tar-like substance called gum.

Chapter 13 – The Diviners

It's common knowledge that early settlers in North America would sometimes use a "divining rod" to find a good location to dig their water wells. It's less known that diviners were also employed to find oil in both Canada and the U.S. beginning in the 1860s.

A diviner was employed when finding the famous King well in Petrolia that triggered the town's 40-year boom. *The Globe* dated January 15, 1867, reported, "it was located by Mr. Kelsey of Buffalo, a celebrated 'oil smeller'." Divining was a common practice in the early oil history of Petrolia, Pennsylvania as well. The men who sold their services as diviners were called "smellers", "witches", "witchers" and sometimes the American term, "doodlebugs" was used.

An excellent description of divining can be found in a rare 1865 book, *The Oil Regions of Pennsylvania*. It is subtitled *Showing Where Petroleum is Found; How It is Obtained, and At What Cost, With Hints For Whom It May Concern*. Below is an excerpt from the text written by William Wright.

"A new profession of men, claiming to be gifted with extraordinary powers, has arisen in Petrolia (Pennsylvania), namely 'oil-smellers' or 'diviners'...No devil, demon, ghost, ghoul, fairy, goblin or table-tapping spirit is known or believed to be at work, albeit the use of a twig of witch hazel or peach might readily enough suggest to some the calling up of spirits from their vasty deep by modern chanters.

"The mode of operating is substantially as follows: The diviner cuts from one of the trees mentioned, a bifurcated bough or twig, reducing the stem and the forks to about a foot in length, for convenience' sake. In each hand, he grasps firmly one end of the fork, letting the stem point upward and a little inward. The hand should

The famous King Well in Petrolia was found with the help of a celebrated "oil smeller".

"...while oil may not be struck where he directs, it is useless to sink where he has pronounced none to exist."

be held with their backs downward. With this simple apparatus, off goes the 'smeller'; and, on arriving above an oil vein, it is claimed that the twig will turn round in his hands, in spite of his utmost exertions, until the stem points directly downward. It may be grasped so tightly that the rind will peel off by the operation; yet this will not prevent the revolution in his hands.

"The author once witnessed this operation going on in the hands of a gentleman of much intelligence and the utmost veracity, who was not a believer in the oil-smeller's claims or pretensions, yet had to acknowledge the existence of the phenomenon for which he could not account.

"It appears that the twig has not this remarkable power in the hands of all persons; for the author was unable to perceive any change or tendency in the wand in his own hand, on arriving at the same spot. Whether the difference were owing to magnetic influence or other cause, is unknown; as also whether the motion betokens the presence of water, petroleum, both, or neither...

"In the oil regions, some of the most productive wells have been located by oil-smellers; in more cases, however, their vaticinations of first class works turn out mere moonshine. However, the diviners have become a power in Petrolia (Pennsylvania), among a people as keenly inquisitive and practical as are to be found, who reason in this way: 'If there is any thing in oil-smelling, we may as well avail ourselves of it as not; for the diviner charges only from twenty five to one hundred dollars for his services in examining a tract; and this is an inconsiderable item in the general expense, seeing we mean to bore any how.'

"There is a pretty general impression that he is a better guide negatively than positively; that while oil may not be struck where he directs, it is useless to sink where he has pronounced none to exist. In a word, the charmer, magnetizer, or natural magician has more real power among the operators than the latter are willing to openly concede."

Col. Harkness, the renowned researcher of Ontario's early oil history, shows even deeper scepticism in his manuscript, *Makers of Petroleum History*. He was able to find an account of a reporter who travelled with an oil diviner to Oil Springs and wrote of this experince in the April 14, 1865 issue of *The Globe*.

It reads in part: "There is a man

here, on competent authority pronounced to be a 'witch' – the masculine wizard being unknown in this locality ...The wizard chooses a forked branch, the end of which he holds tightly in a horizontal position. And then he marches in a stately fashion over the ground, until, by a power to him irresistible, a branch of the stick is drawn down in the direction of the 'ile' vein. When once there, the strongest men in Eniskillen might strive in vain to move it. It will not budge....

"There are some sceptics here who laugh at the 'superstition' as they are pleased to call it; but there are plenty ready to testify to the truth of the facts set forth. They know it, they have seen it, they experience the benefits, and would no more think of commencing a well without first consulting the man witch and his 'crazy' stick than they would of going to bed without a 'nightcap.'

"There is no necessity, however, for people seeking oil anywhere outside Enniskillen to send for the wizard, for your correspondent is firmly convinced that a willow wand will be just as efficacious in the hands of any other man as in those of the Witch of Oil Springs."

Harkness notes that diviners missed rather important discoveries in Oil Springs. Though they were present in the 1860s, they failed to find the oil that created the second boom of Oil Springs in the 1880s. With tongue firmly in cheek, Harkness wrote, "For some inexplicable reason, diviners never become wealthy, although they claim the power of reaching the treasure vaults hidden in the earth; from which one has reason to conclude that after 3,000 and more years, divining is still far from infallible."

Continuing in this droll tone, he added, "All oil diviners known to the writer...have a unique conception of an oil or gas field...These men state that the oil occurs in 'rivers' or 'veins' and is distributed vertically and not horizontally. They maintain (and demonstrate with their device) that the well must be located within a lateral distance of one foot, and that if the drill deviates from the vertical, the 'vein' will be missed. It is scarcely necessary to elaborate on this grave departure from the facts so well known to all who are familiar with the industry, but the writer (an avowed sceptic) cannot refrain from remarking that their theory provides a ready means of explaining or excusing the many dry holes."

"For some inexplicable reason, diviners never become wealthy, although they claim the power of reaching the treasure vaults hidden in the earth."

Chapter 14 – The Fairbank Property in the Elk Hills Oil Lands of California

The story of the Fairbank oil land in Elk Hills, California is a tale that took 56 years from start to finish; a finish that made headlines in Windsor, London, and Sarnia. It began about 1908 when Dr. Charles Fairbank and his partner, Frank Carmen, bought 158 acres of land in Elk Hills from the State of California. They never drilled for oil, but in 1919 they leased it to Standard Oil. Then, as a Canadian Press wire service story reported, "...Out came the gushers and in came the federal government" to claim the land as its own.

Today, this land in California's Kern County is famous for its oil. It's unclear how long Standard Oil continued to make profits and the Fairbank family continued to receive royalties.

The issue of rightful ownership had already been to court several times, but in 1935 the interior secretary Harold Ickes reversed the opinions of his subordinates and made the seizure of land permanent. The government took the land and never reimbursed the Fairbank family.

Robert Fairbank, the younger brother of Charles Sr. who became a U.S. citizen, spearheaded a legal case against the U.S. government. He started his campaign about 1948. The fight would continue for 16 years. He was asking for $15 million in compensation for the confiscated land.

The court case hinged on one issue: whether it was "known" mineral land when it was purchased. Back in 1903, the U.S. federal government gave the land to the State of California but if it was "known mineral land" the federal government was to keep title.

The story of the Elk Hills land concluded in a Washington Court of Claims in January, 1964. Charles Sr., the executor of the Fairbank estate, and Robert, were in the courtroom when it was announced they lost the case. They

"Out came the gushers and in came the federal government" to claim the land as its own.

would not be receiving $15-million; they would not be receiving anything. The case was closed.

The following are two news stories covering the case in more detail:

The Windsor Daily Star, front page, May 9, 1959.

The Fairbank family would not be receiving $15 million; it would not be receiving anything.

Headline: Lambton Family Asks $15 Million from Uncle Sam

Subhead: Fairbank Kin at Petrolia Involved in Oil Dispute

PETROLIA – A "million to one" chance may bring a fortune to a Petrolia family, if the United States Court of Claims upholds their contention in hearings presently in progress at Washington, D.C.

The fortune, $15,000,000 claimed from the United States Government, is not the important issue at stake in the case, according to Robert Fairbank, of Honolulu, and formerly of Petrolia, who is presenting the claim.

Mr. Fairbank said in Washington that the case actually involves primarily a "matter of principle" which he synopsised as "the dangers of administrative law".

The suit is the latest move in a series of legal developments which began in 1935, when the Fairbank family lost title to the 158 acres of oil producing land in the San Joaquin Valley of California, which they had held since 1908.

Mr. Fairbank is one of the four persons primarily interested in the court action. Others are his brothers, Charles O. and Henry Fairbank, both of Petrolia, and a niece, Mrs. William Oakes (Claire Fairbank Oakes), of Toronto, daughter of another brother, John who died some years ago.

At Petrolia today, C.O. Fairbank (Charles Sr.) confirmed the story of the action taken by Robert, who had been an American citizen for years. He said that his brother deserves "all the credit in the world for his perseverance" in keeping the lease alive, long after Standard Oil Co. of California, also involved at one time, backed out of the situation and paid $7,000,000 back royalties to the government.

The case began to develop in 1900, when Dr. C.O. Fairbank, father of three brothers, (editor's note: there were actually four brothers – John Henry Jr., who died in 1927; Charles; Henry and Robert) and a Petrolia oil pioneer, teamed up with the late Frank Carmen to develop oil properties near Bothwell.

Their venture extended to California in 1908, when Carmen, an oil geologist, decided that the San Joaquin Valley could be a potential oilfield and the partners procured 158 acres from the California government.

The land had been ceded to the state by the federal government in the same year, with the restriction that if the land were "known mineral land" it would revert to federal control.

Fairbank and Carmen leased the land to the Standard Oil Co. of California, and in 1916, the combine offered the property to the United States Navy.

Franklin Delaware Roosevelt, then undersecretary, rejected it as worthless. Three years later, exploratory drilling produced oil.

When the Teapot Dome oil scandal broke in the early 1920s, the property in the United States was investigated but it was given a clean bill of health.

Carmen and Dr. Fairbank subsequently died, and the Fairbank family continued ownership of the property until 1935 when Harold Ickes, then Secretary of the Interior, ruled administratively under the land laws that the Fairbank property had been "known oil land" at the time of the sale in 1908 and was therefore federal property.

The Fairbanks and Standard Oil, as well as owners of adjacent lands, were ousted without any form of compensation.

Ickes' action apparently had the sanction of the president, the same Franklin Delano Roosevelt who, 20 years earlier, had characterized the land as "worthless".

The Supreme Court reviewed the matter in 1940, and without reviewing the issue of how anybody could have predicted the oil in 1903, affirmed Mr. Ickes' right and authority to make the decision.

At that time Standard Oil and the others dropped out, with Standard making the $7 million lump payment of back royalties, but the Fairbank interests held on.

Robert Fairbank took no active part in the fight until his discharge from the army in 1947, but at that time, still rankling from the rebuff, and the fact that the case had never been heard on its merits in court, decided to do something about it.

Rep. Charles M. Teague, California Republican, was persuaded to introduce a private bill in 1956. This bill was recommended by the Judiciary Committee, with some criticism of the Icke's decision, and it was passed by Congress.

The bill requested the Court of Claims to rule on the merit of the

Franklin Delaware Roosevelt had earlier deemed the land worthless.

Papers relating to the case fill a whole series of filing cabinets in Petrolia, Honolulu and Washington.

"known oil land" in 1903 claim, and to decide how much, if any, of the Fairbank $15 million claim should be paid. The hearing started May 1.

Charles O. Fairbank said today that the present status of the case is that it is "finally getting its day in court." However, he added, the case may be long drawn-out, lasting for a few months or even years before it is decided.

And, even then, the decision may go against the Fairbank heirs.

Even if there is no financial gain, the Fairbank family has plenty of souvenirs of the oil land issue.

An accumulation of documents, letters, maps and other papers relating to the case fills a whole series of filing cabinets in Petrolia, Honolulu, Washington and other places where they have been filed with family members and lawyers for safekeeping.

As far as the family is concerned, it's still a case of "principle first – cash later, if we get it."

On January 30, 1964 *The Advertiser Topic* in Petrolia carried the following story:

Headline: $15 million suit lost by Fairbanks

The United States Court of Claims has turned down an appeal by Charles Fairbank of Petrolia and his brother, Robert, of Morro Bay, California, for compensation in the expropriation of California oil lands originally owned by their late father.

Charles Fairbank told *The Advertiser Topic* Tuesday he had been informed unofficially of the failure of the $15 million suit.

The legal battles, which have cost the Fairbank family many thousands of dollars, date back 40 years.

The case dates back to 1903 when the U.S. federal government ceded land to California, proceeds from which were to be used for school purposes. If, however, the land was known mineral land, the federal government was to keep title.

Satisfied the land had no hidden riches, California sold it to the late Charles Fairbank, who in 1919 leased it to Standard Oil. The company found oil. Washington stepped in two years later and claimed the land as "known mineral land".

Court battles began, but in 1935 Harold Ickes, secretary of the interior, made the seizure final.

In recent years, due chiefly to the efforts of Robert Fairbank, the case was reopened with a Congressional bill allowing the Fairbanks their day in court.

PART FIVE
Resources and Index

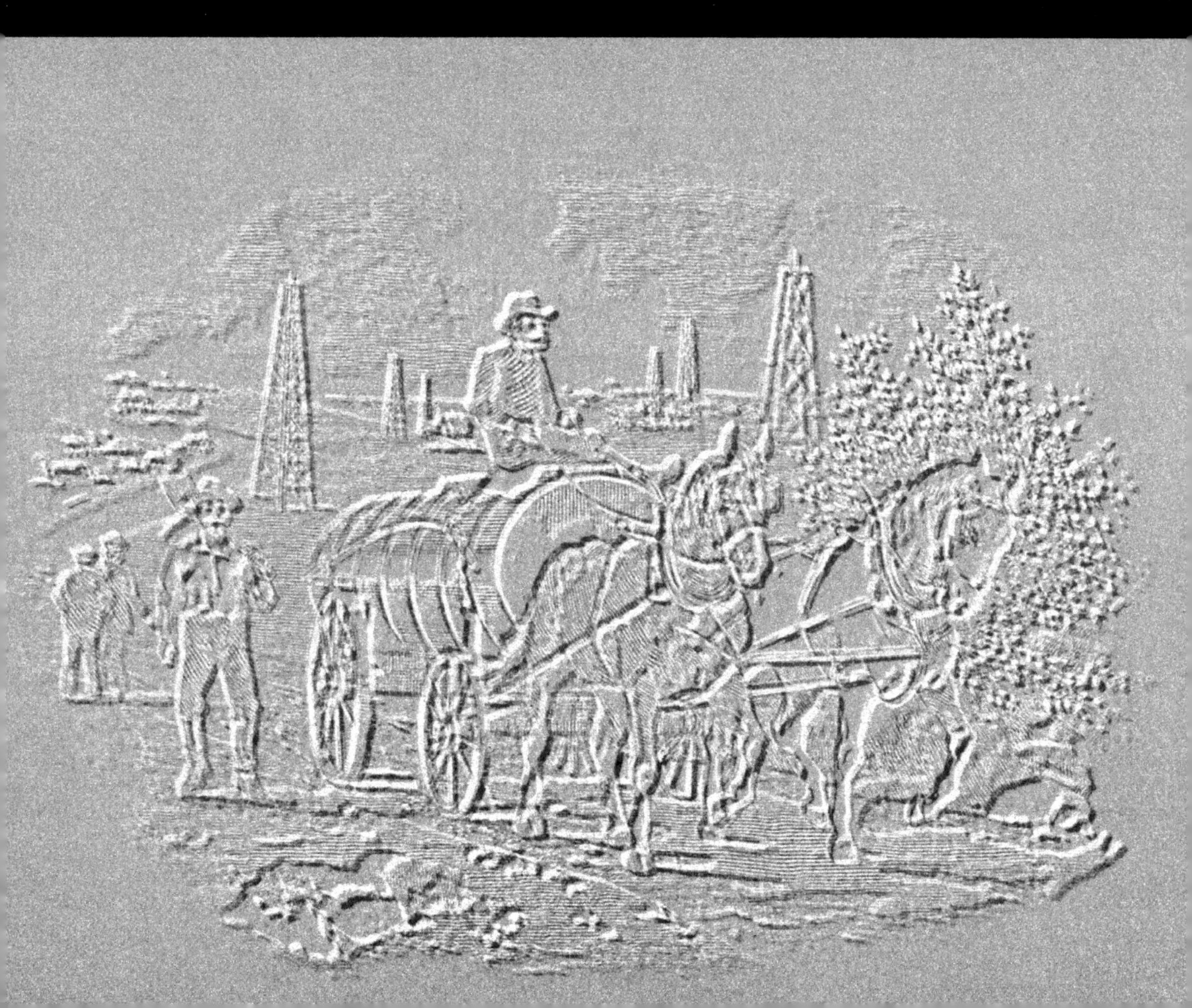

Resources

Unpublished Sources

- Interviews during 2002 and 2003 with Charlie Fairbank, Edward Phelps, Charlie Whipp, Dr. Emory Kemp, Robert Cochrane, Pat Stephen, Ron Burnie, Brad Loosely, Sylvia Fairbank, Bertha Gleeson, Mary Pat Gleeson, Florence Moore
- Speeches by Charlie Fairbank
- Fairbank Letters, Diaries, Documents
- *John Henry Fairbank of Petrolia* (1831-1914) A *Canadian Entrepreneur* Faculty of Graduate Studies, University of Western Ontario, London, Canada By Edward Phelps, 1965
- *Imperial in the Beginning,* The Story of Early Oil History, Before and During Imperial Oil's Early Oil Refinery Operations In and Around the Petrolia Region By Donald Smith, 1989
- *Makers of Oil History,* 1850 *to* 1880, unpublished manuscript By Col. Bruce Harkness
- Oil Museum of Canada, Oil Springs
- Hillsdale Cemetery, Fairbank Mausoleum, Petrolia.

Publications

- *The American Petroleum Industry, the Age of Illumination* 1859-1899 By Harold F. Williamson and Arnold Daum, Northwestern University Press, 1959
- *Belden's Historical Atlas of the County of Lambton, Ontario,* 1880 N.P: Belden, 1881; Reprint Sarnia, Ontario: Edward Phelps, 1973
- *Canada's Tale of Toil & Oil* By Patricia McGee, The Petrolia Discovery Foundation Inc., Stan-McCallum Lithographing, London, Second printing 2000
- *Canada West's Last Frontier,* A *History of Lambton* By Jean Turnbull Elford, Lambton

Historical Society, 1992

- *Canadians At Work*
  By Maynard Hallman, Longmans, Green and Company, Toronto, 1950
- *Canadian Global Almanac* 2001
  MacMillan Canada
- *The Canadian Encyclopedia*
  Hurtig Publishers, Edmonton, Second Edition, 1988
- *Early Development of Oil Technology, Oil Tools and Stories*
  By Wanda Pratt & Phil Morningstar, Sponsored by the Oil Museum of Canada, Browns Graphics and Printing Inc., Petrolia, 1987
- *Equinox Magazine*
  Small-Time Crude
  By David Toole,
  July-August 1986 issue
  Equinox Publishing
- *Gas Lighting*
  By David Gledhill,
  Shire Publications Ltd., United Kingdom, 1987
- *Genealogy of The Fairbanks Family in America* 1633-1897
  By Lorenzo Sayles Fairbanks, American Printing and Engraving Company, Boston, 1897
- *Geological Survey of Canada, Annual Report*, 1890-91
- *The Good Old Days*, The History of Education in Enniskillen Township
  Lawrence A. Crinch, Pole Printing Inc. Forest, Ontario. 1991
- *Hard Oiler!* The Story of Early Canadians' Quest for Oil at Home and Abroad
  By Gary May,
  Dundurn Press, Toronto, 1998
- *A History of the Chemical Industry in Lambton County*
  By R.W. Ford,
  The Sarnia-Lambton Environmental Association, 4th printing, 2001
- *Inventors, Profiles in Canadian Genius*
  By Thomas Carpenter,
  Camden House, Camden East, 1990
- *Lambton County's Hundred Years* 1849 – 1949
  By Victor Lauriston,
  Haines Frontier Printing Company, Sarnia, 1949
- *Lambton's Industrial Heritage*
  By Christopher Andreae,
  Guide Book for Society for Industrial Archaeology, 2000
- *The London and Middlesex Historian*
  The Fenians Are Coming
  By John Mombourquette.
  Autumn 1993 issue
- *Northern Prospects in the 21st Century*, 2000 *Eastern Section*. American Association of Petroleum Geologists.

29th Annual Meeting. Oil Heritage Tour of Lambton County: The Birthplace of the Canadian Oil Industry
By Robert O. Cochrane, Cairnlins Resources Ltd. & Charles Fairbank

- *Oil Springs: The Birthplace of the Oil Industry in North America*
  By Michael O'Meara,
  Oil Springs Centennial Committee, 1958
- *The Party's Over,* Oil, War and the Fate of Industrial Societies
  By Richard Heinberg,
  New Society Publishers, Gabriola Island, B.C., Canada, 2003
- *Petroleum in Canada*
  By Victor Ross, Printer Unknown, 1917
- *Petroleum - Prehistoric to Petrochemicals*
  By G.A. Purdy,
  Copp Clark Publishing Company, Vancouver, Toronto, Montreal, 1957
- *Petrolia* 1866 - 1966
  By Charles Whipp & Edward Phelps, Petrolia, Ontario.
  Petrolia Advertiser Topic, 1966
- *Petrolia,* 1874 – 1974
  By Charles Whipp & Edward Phelps, Petrolia, Print & Litho, 1974
- *Petrolia Canada* 1908, *Souvenir book of the Petrolia Old Boys Reunion,*
  Publisher uhknown
- *Petrolia Old Boys Reunion,* 1946
  printed by Petrolia Topic
- *Petrolia Oil Exchange, The Constitution and Rules,* The Topic Publishing Company, Petrolia, Ont. 1885
- *Report of the Commission Upon The Mineral Resource of Ontario,* 1890
- *Rivers of Oil,* The Founding of North America's Petroleum Industry
  By Hope Morritt,
  Quarry Press, Kingston, 1993
- *Sarnia, A Picture History of the Imperial City*
  By Glen C. Phillips, Iron Gate Publishing Company, Sarnia, 1990
- *Sketches of Creation*
  By Alexander Winchell,
  Harper & Brothers, Publishers, New York, 1872
- *Survivals, Aspects of Industrial Archaeology in Ontario*
  By Dianne Newell and Ralph Greenhill, Boston Mills Press, 1989

Index

C

D

E

F

G

H

I

W